Springer Theses

Recognizing Outstanding Ph.D. Research

Aims and Scope

The series "Springer Theses" brings together a selection of the very best Ph.D. theses from around the world and across the physical sciences. Nominated and endorsed by two recognized specialists, each published volume has been selected for its scientific excellence and the high impact of its contents for the pertinent field of research. For greater accessibility to non-specialists, the published versions include an extended introduction, as well as a foreword by the student's supervisor explaining the special relevance of the work for the field. As a whole, the series will provide a valuable resource both for newcomers to the research fields described, and for other scientists seeking detailed background information on special questions. Finally, it provides an accredited documentation of the valuable contributions made by today's younger generation of scientists.

Theses are accepted into the series by invited nomination only and must fulfill all of the following criteria

- They must be written in good English.
- The topic should fall within the confines of Chemistry, Physics, Earth Sciences, Engineering and related interdisciplinary fields such as Materials, Nanoscience, Chemical Engineering, Complex Systems and Biophysics.
- The work reported in the thesis must represent a significant scientific advance.
- If the thesis includes previously published material, permission to reproduce this must be gained from the respective copyright holder.
- They must have been examined and passed during the 12 months prior to nomination.
- Each thesis should include a foreword by the supervisor outlining the significance of its content.
- The theses should have a clearly defined structure including an introduction accessible to scientists not expert in that particular field.

More information about this series at http://www.springer.com/series/8790

Jordi Naqui

Symmetry Properties in Transmission Lines Loaded with Electrically Small Resonators

Circuit Modeling and Applications

Doctoral Thesis accepted by
Universitat Autònoma de Barcelona, Spain

Springer

Author
Dr. Jordi Naqui
Escola d'Enginyeria, Departament d'Enginyeria Electrònica
Universitat Autònoma de Barcelona
Barcelona
Spain

Supervisor
Prof. Ferran Martín
Escola d'Enginyeria, Departament d'Enginyeria Electrònica
Universitat Autònoma de Barcelona
Barcelona
Spain

ISSN 2190-5053 ISSN 2190-5061 (electronic)
Springer Theses
ISBN 978-3-319-24564-5 ISBN 978-3-319-24566-9 (eBook)
DOI 10.1007/978-3-319-24566-9

Library of Congress Control Number: 2015950878

Springer Cham Heidelberg New York Dordrecht London

Printed on acid-free paper

Springer International Publishing AG Switzerland is part of Springer Science+Business Media
(www.springer.com)

Parts of this thesis have been published in the following articles:

Journals

J. Naqui, A. Fernández-Prieto, F. Mesa, F. Medina, and F. Martín, "Effects of inter-resonator coupling in split ring resonator loaded metamaterial transmission lines," *J. Appl. Phys.*, vol. 115, no. 194903, May 2014.

J. Naqui and F. Martín, "Mechanically reconfigurable microstrip lines loaded with stepped impedance resonators and potential applications," *Int. J. Antennas Propag.*, vol. 2014, Article ID 346838, 8 pages, Feb. 2014.

J. Naqui and F. Martín, "Angular displacement and velocity sensors based on Electric-LC (ELC) loaded microstrip lines," *IEEE Sensors J.*, vol. 14, no. 4, pp. 939–940, Apr. 2014.

J. Naqui, M. Durán-Sindreu, F. Martín, "Selective mode suppression in microstrip differential lines by means of electric-LC (ELC) and magnetic-LC (MLC) resonators," *Appl. Phys. A*, vol. 115, no. 2, pp. 637–643, May 2014.

J. Naqui, and F. Martín, "Transmission lines loaded with bisymmetric resonators and their application to angular displacement and velocity sensors," *IEEE Trans. Microw. Theory Techn.*, vol. 61, no. 12, pp. 4700–4713, Dec. 2013.

J. Naqui, M. Durán-Sindreu, and F. Martín, "Differential and single-ended microstrip lines loaded with slotted magnetic-LC resonators," *Int. J. Antennas Propag.*, vol. 2013, Article ID 640514, 8 pages, May 2013.

J. Naqui, M. Durán-Sindreu, and F. Martín, "Modeling split ring resonator (SRR) and complementary split ring resonator (CSRR) loaded transmission lines exhibiting cross polarization effects," *IEEE Antenn. Wireless Propag. Lett.*, vol. 12, pp. 178–181, Feb. 2013.

J. Naqui, M. Durán-Sindreu, and F. Martín, "Selective mode suppression in coplanar waveguides using metamaterial resonators," *Appl. Phys. A*, vol. 109, no. 4, pp. 1053–1058, Dec. 2012.

J. Naqui, A. Fernández-Prieto, M. Durán-Sindreu, F. Mesa, J. Martel, F. Medina, and F. Martín, "Common mode suppression in microstrip differential lines by means of complementary split ring resonators: theory and applications," *IEEE Trans. Microw. Theory Techn.*, vol. 60, no. 10, pp. 3023–3034, Oct. 2012.

J. Naqui, M. Durán-Sindreu, A. Fernández-Prieto, F. Mesa, F. Medina, and F. Martín, "Multimode propagation and complex waves in CSRR-based transmission line metamaterials," *IEEE Antenn. Wireless Propag. Lett.*, vol. 11, pp. 1024–1027, Aug. 2012.

J. Naqui, M. Durán-Sindreu, and F. Martín, "Alignment and position sensors based on split ring resonators," *Sensors—Special Issue Ultra-Small Sensor Systems and Components*, vol. 12, no. 9, pp. 11790–11797, Aug. 2012.

J. Naqui, M. Durán-Sindreu, J. Bonache and F. Martín, "Implementation of shunt connected series resonators through stepped-impedance shunt stubs: analysis and limitations," *IET Microw. Antennas Propag.*, vol. 5, no. 11, pp. 1336–1342, Aug. 2011.

J. Naqui, M. Durán-Sindreu and F. Martín, "Novel sensors based on the symmetry properties of split ring resonators (SRRs)," *Sensors—Special Issue Metamaterials for Sensing*, vol. 11, no. 8, pp. 7545–7553, Jul. 2011.

Conferences

J. Naqui, L. Su, J. Mata, and F. Martín, "Analysis of transmission lines loaded with pairs of coupled resonant elements and application to sensors," *Moscow Int. Symp. Magn.* (*MISM*-2014), Moscow, Russia, Jun. 2014.

J. Naqui, C. Damm, A. Wiens, R. Jakoby, L. Su, and F. Martín, "Transmission lines loaded with pairs of magnetically coupled stepped impedance resonators (SIRs): modeling and application to microwave sensors," *IEEE MTT-S Int. Microw. Symp.*, Tampa, FL, USA, Jun. 2014.

J. Naqui, M. Durán-Sindreu, and F. Martín, "Strategies for the implementation of sensors and RF barcodes based on transmission lines loaded with symmetric resonators," *21st Int. Conf. on App. Electromagnetics and Communications* (*ICECom 2013*), Dubrovnik, Croatia, Oct. 2013.

J. Naqui, M. Durán-Sindreu, F. Martín, A. Fernández-Prieto, F. Mesa, and F. Medina, "Complex modes in periodic transmission lines based on split rings," *International Conference on Electromagnetics in Advanced Applications* (*ICEAA13*), Torino, Italy, Sept. 2013.

J. Naqui, M. Durán-Sindreu, F. Martín, "Transmission lines loaded with bisymmetric resonators and applications," *IEEE MTT-S Int. Microw. Symp.*, Seattle, WA, USA, Jun. 2013.

J. Naqui, M. Durán-Sindreu, and F. Martín, "Selective mode suppression in microstrip differential lines by means of electric-LC (ELC) and magnetic-LC (MLC) resonators," *META 2013: 4th International Conference on Metamaterials, Photonic Crystals and Plasmonics*, Sharjah, UAE, Mar. 2013.

J. Naqui, M. Durán-Sindreu, and F. Martín, "On the symmetry properties of coplanar waveguides loaded with symmetric resonators: analysis and potential applications," *IEEE MTT-S Int. Microw. Symp.*, Montreal, Canada, Jun. 2012.

J. Naqui, M. Durán-Sindreu, and F. Martín, "Selective mode suppression in coplanar waveguides using metamaterial resonators," *META 2012: 3rd International Conference on Metamaterials, Photonic Crystals and Plasmonics*, Paris, France, Apr. 2012.

J. Naqui, M. Durán-Sindreu, A. Fernández-Prieto, F. Mesa, F. Medina, and F. Martín, "Differential transmission lines loaded with split ring resonators (SRRs) and complementary split ring resonators (CSRRs)," *Metamaterials'2011: the 5th Int. Cong. Adv. Electromagn. Materials Microw. Opt.*, Barcelona, Spain, Oct. 2011.

J. Naqui, A. Fernández-Prieto, M. Durán-Sindreu, J. Selga, F. Medina, F. Mesa, and F. Martín, "Split rings-based differential transmission lines with common-mode suppression," *IEEE MTT-S Int. Microw. Symp.*, Baltimore, MD, USA, Jun. 2011.

J. Naqui, M. Durán-Sindreu, J. Bonache and F. Martín, "Stepped impedance shunt stubs (SISS): analysis and potential applications in planar materials," *4th Young Scientist Meeting Metamaterials (YSMM11)*, Valencia, Spain, Feb. 2011.

Workshops

J. Naqui, A. K Horestani, C. Fumeaux, and F. Martín, "Microwave sensors based on symmetry properties of resonator-loaded lines," in workshop "Using symmetry-related electromagnetic properties for microwave device design: application to half-mode circuits, balanced lines and circuits, and microwave sensors," *Europ. Microw. Week*, Paris, France, Sept. 2015.

J. Naqui, J. Bonache and F. Martín, "Revisiting equivalent circuit models for emerging technologies: from microwaves to THz," in workshop "Equivalent circuit models for metamaterial-inspired planar circuits based on split rings and related resonators," *IEEE MTT-S Int. Microw. Symp.*, Tampa, FL, USA, Jun. 2014.

J. Naqui, M. Durán-Sindreu and F. Martín, "On the symmetry properties of transmission lines loaded with metamaterial resonators: theory and applications," in workshop "Microwave Metamaterial Concepts, Circuits and Applications," *Europ. Microw. Conf.*, Nurember, Germany, Oct. 2013.

F. Martín, J. Naqui, P. Vélez, M. Durán-Sindreu and J. Bonache, "Split ring resonator (SRR) and stepped impedance resonator (SIR) based metamaterial transmsision lines: application to microwave components and novel sensing strategies," in workshop "Metamaterials in communications and sensing: reality or fiction?," *IEEE Radio Wireless Week*, Austin, TX, USA, Jan. 2013.

Supervisor's Foreword

It is a great pleasure to introduce Dr. Jordi Naqui's thesis work, accepted for publication within Springer Theses and awarded with a prize for outstanding original work. Dr. Naqui joined my Research Group and CIMITEC (the Technology Transfer Centre I head at the Universitat Autonoma de Barcelona) in September 2009. He has had a national predoctoral fellowship, granted by the Spanish Ministry of Education, Culture, and Sports (ref. AP2010-0431), for the development of his research activities during these years that have concluded with the oral presentation and defense of the Ph.D. thesis on October 17, 2014.

Dr. Naqui's thesis includes significant original scientific contributions, representing considerable advance in the field of Microwave Engineering, specifically on the topics of wireless communications, sensing, and radiofrequency identification (RFID). Such contributions have been published or will be shortly published in top journals, international well-recognized conferences, and book chapters. I would like to emphasize that the work carried out by Dr. Naqui, reported in this book, is very extensive and complete. It covers the research cycle in the area of experimental sciences and engineering, including fundamentals/theory, design/simulation, and fabrication/characterization, and a wide variety of proof-of-concept demonstrators of the novel ideas generated in the thesis are reported.

In the thesis, a new design principle, based on the symmetry-related electromagnetic properties of transmission lines loaded with electrically small and symmetric resonators (mostly inspired by metamaterials), is introduced for the first time and is transversely applied to the different types of devices considered. This design principle is the most important contribution of the thesis. It is described in detail in the thesis and the potentiality of this novel design approach is demonstrated through the development of a wide variety of proof-of-concept devices, including common-mode suppressed differential lines and balanced filters, microwave sensors (for linear/angular displacement and rotation speed measurements, for alignment purposes and for dielectric characterization), and spectral signature barcodes. Thus the thesis clearly identifies different scenarios where the proposed design principle can be applied. I would like to highlight the novel sensing concept reported in the thesis, based on symmetry disruption, hence resulting in microwave sensors robust against changes in environmental conditions (the reported contactless angular displacement and velocity sensors are good examples of such sensors).

It is also remarkable that the different structures considered in this work, based on resonator-loaded lines, have been modeled by means of lumped element equivalent circuits, very useful for design purposes. These models have been validated through parameter extraction methods and comparison to electromagnetic simulations and experimental results. Moreover, the thesis includes the modeling of two important effects in resonator-loaded lines not considered so far: (*i*) the arbitrary orientation of split rings in lines loaded with such elements, where mixed (magnetoelectric) coupling between the line and the rings arises and (*ii*) the modeling of transmission lines with inter-resonator coupling, where stopband bandwidth enhancement is explained by the appearance of complex modes.

Some parts of the research activity reported in this book have been carried out in collaboration with other Groups, particularly, with the Group of Prof. Francisco Medina (University of Seville) on the topic of common-mode suppressed balanced lines and filters, with the Group of Prof. Christophe Fumeaux (University of Adelaide) in the field of spatial sensors, and with the Group of Prof. Rolf Jakoby (Darmstadt University of Technology) on the topic of sensors for dielectric characterization (Dr. Naqui made a fruitful three-month stay in Darmstadt before the Ph.D. defense). Such collaborations have opened new research horizons in the near future, and Dr. Naqui's thesis constitutes the seed for new research lines within my Group, which will represent novel Ph.D.'s in a few years.

And last but not least, it is remarkable that thanks to the research activity carried out in this thesis, a contract with the European Space Agency in collaboration with the company *EMXYS Embedded Instruments*, focused on the development of angular velocity sensors for reaction wheels, has been achieved. Within the topic of RFID, printed radiofrequency barcodes (i.e., chipless RFID tags operating in the frequency domain) will be developed in the framework of a collaborative project with the company *Scytl*, for electronic voting and tracking of election processes. These are two examples of the potential of the ideas generated in this thesis. Dr. Naqui is a key researcher in these projects, and a large part of his time is now dedicated to them. Nevertheless, after the Ph.D. presentation he has given continuity to the research activities, trying to bring beyond the state-of-the-art ideas and concepts of interest for the academia and useful to the industry. Being aware of the passion and enthusiasm of Dr. Naqui for Research, I hope to see him involved in the university for many years.

Bellaterra, Barcelona
April 2015

Prof. Ferran Martín

Acknowledgments

I would like to express my most sincere gratitude to those who have contributed to the present thesis.

First of all, many thanks to my supervisor Ferran Martín, who gave me the opportunity to begin the Ph.D. His knowledge, motivation, and continuous help during these past years have played a key role in this thesis. I would also like to thank the people from the research center CIMITEC, the Department of Electronics Engineering, and the Engineering School at the Universitat Autònoma de Barcelona. Among them, I am particularly very grateful to Gerard Sisó and Javier Hellín. Although they are not formally coauthors of the publications within the context of this thesis, they are of course part of them. Almost all experimental results in this thesis are merits of their outstanding and patient work on fabrication-related tasks.

I am also very grateful to my colleagues from national and international universities with whom I have collaborated. Many thanks to Francisco Medina, Francisco Mesa, and Armando Fernández from University of Seville for our continuous collaboration. Many thanks to Rolf Jakoby and Christian Damm to let me carry out a great stay in Darmstadt University of Technology.

Many thanks to my friends, especially to Marín, Novo, Dani, and Núria. Thanks also to my pet Dana, my Boxer dog, for receiving me with so much enthusiasm every weekday. Lastly, special gratitude to those who have really suffered my difficult and stressful moments, my family. Especially, my wife Amalia and my little daughter Martina, my mother Mercè, my grandmother Mercè, and my parents-in-law Antonio and Amalia. Thank you for your patience.

Jordi Naqui

Acknowledgments

Contents

1 Introduction 1
1.1 Motivations 1
1.2 Organization 2
1.3 Funding 3

2 Fundamentals of Planar Metamaterials and Subwavelength Resonators 5
2.1 Electromagnetic Metamaterials 6
2.1.1 Material Classification 7
2.1.2 Left-Handed Media 7
2.2 Transmission-Line Metamaterials 10
2.2.1 Application of the Transmission-Line Theory to Metamaterials 10
2.2.2 Composite Right-/Left-Handed (CRLH) Transmission Lines 13
2.2.3 *CL*-Loaded and Resonant-Type Approaches 16
2.2.4 Resonant-Type Single-Negative Transmission Lines 18
2.2.5 Discussion About Homogeneity and Periodicity 23
2.3 Metamaterial-Based Resonators 24
2.3.1 Split-Ring Resonator (SRR) 25
2.3.2 Double-Slit Split-Ring Resonator (DS-SRR) 29
2.3.3 Folded Stepped-Impedance Resonator (FSIR) 30
2.3.4 Electric Inductive-Capacitive (ELC) Resonator 31
2.3.5 Complementary Resonators 32
2.4 Magneto- and Electro-Inductive Waves 36
2.4.1 Magneto-Inductive Waves in Arrays of Magnetically-Coupled Resonators 36
2.4.2 Electro-Inductive Waves in Arrays of Electrically-Coupled Resonators 39
References 40

3 Advances in Equivalent Circuit Models of Resonator-Loaded Transmission Lines . . . 45
3.1 Line-to-Resonator Magnetoelectric Coupling . . . 46
3.1.1 Coplanar Waveguides Loaded with Pairs of SRRs and CSRR-Loaded Microstrip Lines . . . 46
3.2 Inter-Unit-Cell Inter-Resonator Coupling . . . 52
3.2.1 Coplanar Waveguides Loaded with Pairs of SRRs and CSRR-Loaded Microstrip Lines . . . 52
3.3 Limits on the Synthesis of Electrically Small Resonators . . . 62
3.3.1 Microstrip Stepped-Impedance Shunt-Stubs (SISSs) . . . 63
References . . . 69

4 On the Symmetry Properties of Resonator-Loaded Transmission Lines . . . 73
4.1 On the Symmetry Properties of Transmission Lines . . . 73
4.2 On the Alignment of Symmetry Planes . . . 76
4.2.1 SRR- and CSRR-Loaded Coplanar Waveguides . . . 79
4.2.2 SRR- and CSRR-Loaded Differential Microstrip Lines . . . 85
4.2.3 ELC- and MLC-Loaded Differential Microstrip Lines . . . 88
4.3 On the Misalignment of Symmetry Planes . . . 95
4.3.1 SRR- and FSIR-Loaded Coplanar Waveguides . . . 97
4.3.2 SIR-Loaded Microstrip Lines . . . 101
4.3.3 ELC-Loaded Coplanar Waveguides . . . 102
4.3.4 MLC-Loaded Microstrip Lines . . . 113
4.4 On the Generalization of Symmetry Rupture . . . 115
4.4.1 Microstrip Lines Loaded with Pairs of SISSs . . . 118
4.4.2 Coplanar Waveguides Loaded with Pairs of SRRs . . . 121
References . . . 122

5 Application of Symmetry Properties to Common-Mode Suppressed Differential Transmission Lines . . . 125
5.1 Introduction . . . 125
5.2 Symmetry-Based Selective Mode Suppression . . . 127
5.3 Common-Mode Suppressed Differential Microstrip Lines . . . 127
5.3.1 CSRR- and DS-CSRR-Loaded Differential Microstrip Lines . . . 127
5.3.2 ELC- and MLC-Loaded Differential Microstrip Lines . . . 140
References . . . 140

6 Application of Symmetry Properties to Microwave Sensors . . . 143
6.1 Introduction . . . 143
6.2 Symmetry-Based Sensing . . . 145
6.2.1 Coupling-Modulated Resonance . . . 145
6.2.2 Resonance Frequency Splitting/Shifting . . . 147

6.3 Linear Alignment and Displacement 148
6.3.1 SRR- and FSIR-Loaded Coplanar Waveguides 149
6.3.2 SIR-Loaded Microstrip Lines 152
6.4 Linear Position 155
6.4.1 Coplanar Waveguides Loaded with Split Rings. 155
6.5 Angular Alignment and Displacement 159
6.5.1 SRR- and FSIR-Loaded Coplanar Waveguides 160
6.5.2 ELC-Loaded Coplanar Waveguides. 160
6.5.3 ELC-Loaded Microstrip Lines 164
6.6 Angular Velocity. 166
6.6.1 ELC-Loaded Coplanar Waveguides and Microstrip Lines 167
6.7 Differential Permittivity 170
6.7.1 Microstrip Lines and Conductor-Backed Coplanar Waveguides Loaded with Pairs of SISSs 170
References 174

7 Application of Symmetry Properties to Spectral Signature Barcodes 177
7.1 Introduction 177
7.2 Symmetry-Based Encoding. 180
7.3 Coplanar Waveguide Implementations 182
7.3.1 SRR- and FSIR-Loaded Coplanar Waveguides 182
7.3.2 ELC-Loaded Coplanar Waveguides. 183
7.4 Microstrip Line Implementations 184
7.4.1 FSIR-Loaded Microstrip Lines 184
7.4.2 MLC-Loaded Microstrip Lines 184
References 186

8 Conclusions and Future Work 189
8.1 General Conclusions 189
8.2 Line-to-Resonator Magnetoelectric Coupling. 190
8.3 Inter-Unit-Cell Inter-Resonator Coupling 191
8.4 Synthesis of Electrically Small Resonators 191
8.5 Symmetry Properties 191
8.6 Common-Mode Suppressed Differential Lines. 193
8.7 Microwave Sensors 193
8.8 Spectral Signature Barcodes 194
8.9 Future Work. 194

Appendix A: Dispersion Relation of Two-Port Networks. 197

Appendix B: Equivalent Circuit Models of Electric and Magnetic Couplings 199

Appendix C: Even/Odd-Mode *S* Parameters 201

Appendix D: Mixed-Mode *S* Parameters 203

Appendix E: Rejection Bandwidth Definition 205

Appendix F: Envelope Detector Design 209

About the Author

Jordi Naqui was born in Granollers, Spain, in 1984. He received the Telecommunication Technical Engineering diploma (specialty in Electronic Systems), the Telecommunication Engineering degree, the Master's degree in Microelectronics and Nanoelectronics Engineering, and the Ph.D. degree in Electronics Engineering from the Universitat Autònoma de Barcelona (UAB), Bellaterra (Barcelona), Spain, in 2006, 2010, 2011, and 2014, respectively. From 2007 to 2009, he conducted internships in Mier Comunicaciones and Ficosa International (Advanced Communications), and he worked as a telecommunication engineering consultant in Sayós & Carrera. He carried out a three-month predoctoral research stay at the Institute for Microwave Engineering and Photonics (IMP) from Darmstadt University of Technology (TUD). He was a recipient of a national predoctoral teaching and research fellowship by the Spanish Government (2012–2014). He received the first prize of Yarman-Carlin Best Student Paper Contest at 14th Mediterranean Microwave Symposium in 2014 and the Young Scientist Award at XXX International Union of Radio Science (URSI) Spanish National Symposium in 2015. He was the co-recipient of IX Comerma Engineer Research Award of Industrial Engineering awarded by the Local Government of Ferrol and University of A Coruña, Spain, in 2015. He has been teaching at the UAB in the Department of Electronics Engineering since 2010. Currently, he is a postdoctoral researcher of CIMITEC at the UAB. His research activities are focused mainly on modeling and design of passive RF/microwave resonant structures, applied to filters, sensors, and chipless Radio-Frequency Identification (RFID) tags. Dr. Naqui is member of the Institute of Electrical and Electronics Engineers (IEEE), and he serves as a reviewer in many international journals.

Acronyms

1D	One-dimensional
2D	Two-dimensional
3D	Three-dimensional
AM	Amplitude modulation
ASM	Aggressive space mapping
BW	Backward wave
CB-CPW	Conductor-backed coplanar waveguide
CMRR	Common-mode rejection ratio
CPS	Coplanar strips
CPW	Coplanar waveguide
CRLH	Composite right/left-handed
CSRR	Complementary split-ring resonator
CSSRR	Complementary split squared ring resonator
DGS	Defected ground structure
DNG	Double-negative
DPS	Double-positive
DS-CSRR	Double-slit complementary split-ring resonator
DS-SRR	Double-slit split-ring resonator
EBG	Electromagnetic bandgap
EIW	Electro-inductive wave
ELC	Electric inductive-capacitive
EM	Electromagnetic
EMI	Electromagnetic interference
ENG	Epsilon-negative
FBW	Fractional bandwidth
FSIR	Folded stepped-impedance resonator
IC	Integrated circuit
LH	Left-handed
LTCC	Low temperature co-fired ceramic
MIW	Magneto-inductive wave
MLC	Magnetic inductive-capacitive
MNG	Mu-negative

NB-CSRR	Non-bianisotropic complementary split-ring resonator
NB-SRR	Non-bianisotropic split-ring resonator
NRI	Negative-refractive index
OCSRR	Open complementary split-ring resonator
OSRR	Open split-ring resonator
PCB	Printed circuit board
PGS	Patterned ground structure
PLH	Purely left-handed
PRH	Purely right-handed
RF	Radio frequency
RFID	Radio-frequency identification
RH	Right-handed
SIR	Stepped-impedance resonator
SISS	Stepped-impedance shunt-stub
SNG	Single-negative
SRR	Split-ring resonator
SSRR	Split squared ring resonator
TEM	Transverse electromagnetic
VNA	Vector network analyzer

Chapter 1
Introduction

1.1 Motivations

Wireless communications, sensors and high-speed digital circuits are increasingly demanding for low-cost solutions, miniaturization, high performance, reconfigurability, or new functionalities. In recent years, novel and potential ideas have emerged in the microwave community from metamaterial concepts for the design of RF/microwave components, circuits, and systems. Typically, metamaterial-based devices are composed of transmission lines loaded with reactive elements (inductors, capacitors, and/or both of them in the form of resonators). These structures have been extensively shown to exhibit high controllability in their passband characteristics. Since the characteristic impedance and the electrical length of these structures are controllable (obviously, to some extent), such a design is referred to as *dispersion and impedance engineering*.

This thesis is aimed at the analysis, circuit modeling, and applications of transmission lines loaded with electrically small resonators (i.e. without individual inductors and capacitors) in planar technology. In contrast to the dispersion and impedance engineering, the controllability in this book is focused on stopband characteristics provided by the resonant elements. Specifically, the main contribution in this book is the study of symmetry-related electromagnetic properties in transmission lines loaded with metamaterial resonators. Since the content in this book expands the state-of-the-art in the field of metamaterial-based and resonator-loaded transmission lines, a review of the fundamental theory of metamaterials and electrically small resonators is firstly provided.

The electromagnetic behavior of resonator-loaded transmission lines in terms of symmetry properties is exploited with an eye to designing innovative RF/microwave devices. Particularly, the theory behind such symmetry properties is applied for filtering, sensing, and encoding purposes. Common-mode suppressed differential lines, novel microwave sensors for spatial variables (alignment, displacement, position and velocity) and differential permittivity, and spectral signature-based chipless

J. Naqui, *Symmetry Properties in Transmission Lines Loaded with Electrically Small Resonators*, Springer Theses,
DOI 10.1007/978-3-319-24566-9_1

radio-frequency identification (RFID) tags, are synthesized in this thesis using the theory derived from symmetry properties. The presented common-mode filters are seen to be competitive in performance as compared to the state-of-the-art, exhibiting simultaneously an efficient wideband common-mode suppression within a broadband transmission for differential-mode signals with low insertion loss. The main advantage of the reported sensors inspired by symmetry-related properties is their inherent robustness against variable environmental factors. The key aspect in the proposed frequency-domain barcodes is that a new alternative encoding approach is reported, which has a new degree of freedom that brings more flexibility in the encoding as compared to the standard strategy.

1.2 Organization

The outline of this thesis is as follows.

- Chapter 1 is the present introduction to the thesis which is aimed at the motivations, organization, and funding projects and grants.
- Chapter 2 is addressed to the fundamental theory of planar metamaterials and electrically small resonators. An important amount of background necessary to understand the thesis is reviewed.
- Chapter 3 presents advances in the lumped-element equivalent-circuit models of transmission lines loaded with electrically small resonators. The mixed magnetoelectric coupling in the line-to-resonator interaction is modeled, the effects of inter-resonator coupling in chains of resonators are investigated, and the limits on the synthesis of shunt-connected series resonators by typical planar electrically small resonators are inferred.
- Chapter 4 analyzes and models the behavior of resonator-loaded transmission lines subjected to symmetry properties. The achieved stopband controllability (frequency, bandwidth, and/or attenuation) is explained in terms of the symmetry/asymmetry of the structures. The novel reported equivalent circuits are in good agreement with the investigated symmetry-related properties.
- Chapter 5 is devoted to the synthesis of common-mode suppressed differential microstrip lines. Although differential signaling allows for high-quality communications, for which balanced (symmetric) structures are required, common-mode noise rejection is usually needed.
- Chapter 6 presents novel sensors whose principle of operation is based on symmetry disruption. Most of the proposed devices are spatial sensors involving the measurement of a spatial physical magnitude, such as displacements. The exception is a preliminary differential comparator and sensor for permittivity measurements.
- Chapter 7 proposes an alternative encoding mechanism based on symmetric/asymmetric configurations for the design of spectral signature-based chipless RFID tags.

- Chapter 8 summarizes the most relevant results and contributions reported in this thesis. Future work on several research lines that have emerged from this thesis is proposed, with special emphasis on technology transfer projects.

As is apparent, this thesis can be classified into two parts. Chapters 1–4 aims at theoretical content and should be read sequentially. On the other hand, Chaps. 5–7 are addressed to applications based on the content explained in the previous chapters, and the order of reading is therefore irrelevant.

1.3 Funding

The work conducted during the realization of this thesis was carried out within the group GEMMA (Grup d´Enginyeria de Microones, Metamaterials i Antenes) and the research center CIMITEC (Centre d´Investigació en Metamaterials per a la Innovació en Tecnologies Electrònica i de Comunicacions), which is part of the Electronics Engineering Department at the Universitat Autònoma de Barcelona.

GEMMA has been recognized as a Consolidated Group by the Catalan Government (AGAUR). CIMITEC is one of the centers of the Technological Innovation Network of TECNIO (ACC1Ó) of the Catalan Government, created with the objective of promoting the transference of technology to industry in the field of Information and Communication Technology. Furthermore, GEMMA/CIMITEC has been part of the European Network of Excellence NoE 500252-2 METAMORPHOSE (Metamaterials organized for radio, millimeter wave and photonic super lattice engineering), the main objective of which was to research, study and promote artificial electromagnetic materials and metamaterials within the European Union. Recently, it has given rise to the Virtual Institute for Artificial Electromagnetic Materials and Metamaterials (METAMORPHOSE VI AISBL).

Among the projects with the different institutions and companies that have given support to the developed research activities or that have emerged from them, it should be highlighted:

- Project RTC-2014-2550-7 Retos de Colaboración 2014 by the Spanish Government granted to a consortium composed of Scytl Secure Electronic Voting, Instituto de Microelectrónica de Barcelona-CNM (CSIC), and UAB-CIMITEC. Title: *Technological development of chipless RFID tags by printing technologies on low-cost substrates for their integration in systems of electronic voting (Chipless RFID).*
- Project TEC2013-40600-R by the Spanish Government. Title: *Design and synthesis of RF/microwave components based on advanced concepts and their application to communication circuits, sensors and RFID.*
- Project 4000111799/14/NL/SC Innovation Triangle Initiative (ITI) by the European Space Agency (ESA) granted to EMXYS Embedded Instruments & Systems, UAB-CIMITEC, and Universitat Politècnica de València (UPV). Title: *Coplanar symmetry-based contactless rotation sensor.*

- Project TEC2011-13615-E EXPLORA by the Spanish Government. Title: *New sensors and detectors based on the symmetry properties of split ring resonators.*
- Project TEC2010-17512 METATRANSFER by the Spanish Government. Title: *New strategies of design and synthesis of microwave components based on metamaterial concepts oriented toward technology transfer.*
- Project CSD2008-00066 CONSOLIDER INGENIO 2010 granted to a consortium composed of eight research groups from different Spanish Universities and funded by the Spanish Government. Title: *Metamaterial engineering (EMET).*

This Ph.D. thesis has also been supported by a national predoctoral fellowship (grant AP2010-0431) from the Training of University Staff Program (FPU) of the Spanish Ministry of Education, Culture, and Sports (MECD).

Chapter 2
Fundamentals of Planar Metamaterials and Subwavelength Resonators

Electromagnetic metamaterials have been a subject of huge research worldwide in the electromagnetic and microwave communities since the first so-called left-handed structure was experimentally conceived in 2000. These efforts have been primarily motivated by the exotic (unusual) electromagnetic properties that these artificial media may feature, such as backward wave propagation. Meanwhile, electrically small resonators (ESRs) based on split rings, which are typical key building blocks to the synthesis of metamaterials, have been investigated broadly as well. The research activity on these resonant particles has been focused not only on the implementation of metamaterials, but also on the design of RF/microwave circuits and components. Thus, novel ideas have emerged in the microwave community to envisage devices inspired by metamaterial concepts with high performance, small size and/or new functionalities. In this regard, in the present thesis these metamaterial-based particles are not utilized to construct metamaterial structures, but to design innovative RF/microwave devices. Nonetheless, a brief introduction to metamaterials is provided for a historical standpoint and to make the thesis more complete, although reporting the current state-of-the-art is beyond the scope of this thesis.

This chapter is devoted to planar metamaterials and electrically small split-ring resonators, mainly from a circuit approach perspective. It should be mentioned that a significant part of the intensive work in such topics has been published in textbooks by several pioneering research groups in the last decade [1–6]. Section 2.1 provides a brief description of the fundamental properties of metamaterials as well as a historical overview. In Sect. 2.2, the different approaches of transmission-line metamaterials are introduced, which are of particular interest to the microwave community. Specifically, the main focus is aimed at the so-called resonant-type approach in planar technology where subwavelength resonators are used for its synthesis. Inside this category, structures with stopband characteristics will receive the most attention. Next, Sect. 2.3 is aimed at describing subwavelength resonators that are used throughout this thesis, with special emphasis on the properties derived from

J. Naqui, *Symmetry Properties in Transmission Lines Loaded with Electrically Small Resonators*, Springer Theses,
DOI 10.1007/978-3-319-24566-9_2

symmetry assumptions. Finally, Sect. 2.4 is addressed to waves induced in chains of coupled resonators, namely, electro-inductive and magneto-inductive waves.

2.1 Electromagnetic Metamaterials

Electromagnetic metamaterials may be broadly defined as *artificial* (man-made) *effectively homogeneous structures* (usually made up of an arrangement of metals and dielectrics) *with unusual properties not readily available in nature* [1].[1] Certainly, the prefix *meta* means *beyond* or *after* in Greek, suggesting that the medium possesses properties that transcend those available in natural materials, this being the most relevant characteristic of metamaterials.[2]

A medium is regarded to be effectively (on average) homogeneous (uniform) if its unit-cell dimension is much smaller than the guided wavelength. This homogenization (averaging) procedure is the key to considering the structure electromagnetically homogeneous along the direction of propagation. Consequently, a metamaterial behaves as a real material in the sense that the macroscopic (in a piece of matter) response to an electromagnetic field can be characterized by effective macroscopic constitutive parameters, namely, the permittivity ϵ and the permeability μ [1, 2]. Indeed, this approach is a direct translation of the characterization of natural media, which are made of atoms and molecules with dimensions many orders of magnitude smaller than the wavelength; the original objective in defining a permittivity and a permeability was to present an homogeneous view of the electromagnetic properties of a discrete natural medium. Analogously, metamaterials are usually synthesized by embedding a periodic[3] array of artificial small inclusions (man-made atoms and molecules) at a sufficiently electrically small mutual distance in a specified host medium [2]. Therefore, the design parameters in the synthesis process of a metamaterial (such as the shape, arrangement or alignment of the inclusions) provide a large collection of independent parameters (or freedom degrees) in order to engineer an artificial material with a specific controllable electromagnetic response different from that obtained by its constitutive materials [2]. In summary, the macroscopic view of metamaterials consists of replacing an artificial structurally inhomogeneous structure with an effectively homogeneous medium characterized by effective constitutive parameters.

---

[1] Since the international scientific community has not achieved a consensus in the definition of electromagnetic metamaterials [6], other artificial inhomogeneous structures with controllable electromagnetic properties (e.g. electromagnetic bandgaps—EBGs—which are based on the Bragg regime) are also sometimes regarded as metamaterials [2].

[2] The subject has been known for a long time as artificial dielectrics, composite materials, or microstructured materials. The aim has always been to reproduce physical responses of known materials or to obtain some desirable responses not readily available in nature.

[3] Periodicity is not fundamental, but eases the design.

2.1.1 Material Classification

Since the response of a medium to an electromagnetic field is determined by its properties (ϵ and μ),[4] this allows for the classification of a medium depending on the sign of the constitutive parameters, as illustrated in Fig. 2.1 [1, 2]. Media with both permittivity and permeability greater than zero are known as *double-positive* (DPS) media. A medium with either permittivity or permeability less than zero is designated as *epsilon-negative* (ENG) medium or *mu-negative* (MNG) medium, respectively, or alternatively *single-negative* (SNG) medium in both scenarios. More interestingly, media with negative permittivity and permeability are called *double-negative* (DNG) media, following the preceding nomenclature. Nevertheless, several other terminologies have been suggested, such as *left-handed* (LH), *backward-wave* (BW), and *negative-refractive index* (NRI) media [1, 2].[5] While natural DPS and SNG media are known to exist (see Fig. 2.1), a DNG medium has not been found in nature to date [1, 2], and for this reason the most popular metamaterials in the considered classification belong to this category [1]. As will be explained in Sect. 2.1.2, DNG media may be synthesized artificially by pairs of elements, in which an array of one of the elements produces a negative effective permittivity and an array of the other element induces a negative effective permeability.

As is well known, the propagation of electromagnetic waves through a medium at the macroscopic level is governed by Maxwell's equations, and such a propagation depends on the constitutive parameters as follows. When either permittivity or permeability is negative, the medium supports only non-propagating evanescent modes (the medium is opaque to signal transmission). Consequently, there cannot be energy transfer through the medium because the incident electromagnetic waves suffer reactive attenuation (reflection, but not dissipation[6]) and decay exponentially in amplitude away from the source [4, 7]. By contrast, when permittivity and permeability are both either positive or negative, wave propagation inside a medium is possible (the medium is said to be transparent).

2.1.2 Left-Handed Media

Left-handedness expresses the fact that the propagation of plane waves is described by the electric field **E**, the magnetic field **H**, and the wave vector or propagation constant **k** building a left-handed triplet, in contrast to conventional materials (with positive constitutive parameters) where this triplet is right-handed (see Fig. 2.2). However, **E**, **H**, and the Poynting vector **S** maintain a right-handed relationship. As a consequence, the phase velocity (parallel to the direction of phase propagation or

---

[4]If losses are absent, as is assumed throughout this chapter, ϵ and μ are real numbers.

[5]The term NRI describes the particularity that the index of refraction in these media is negative, while the terminologies LH and BW will be addressed shortly.

[6]Recall that losses are precluded in the whole chapter.

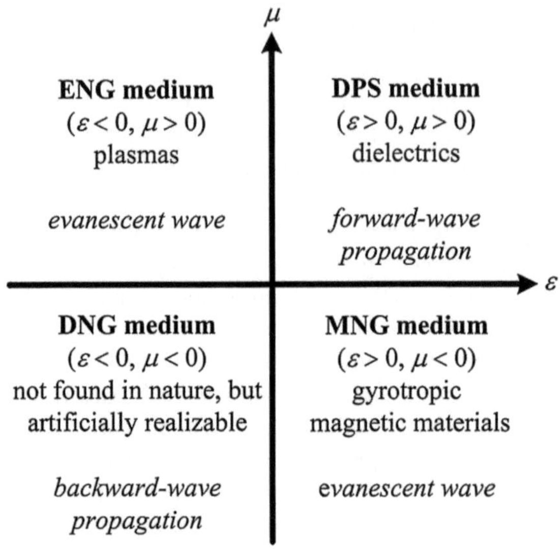

Fig. 2.1 Material classification according to the sign of permittivity-permeability

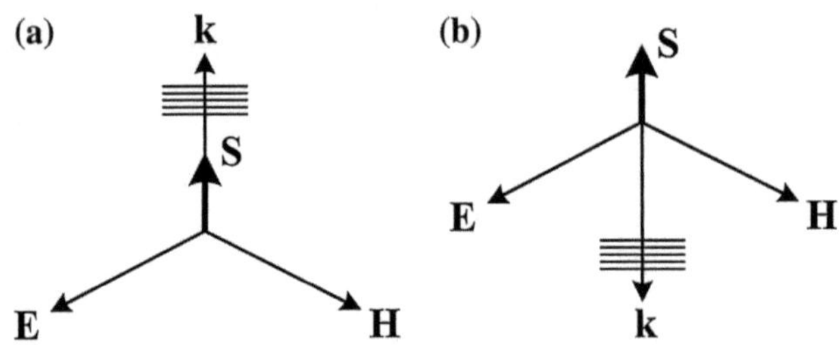

Fig. 2.2 Orientation of the electric field **E**, magnetic field **H**, propagation constant **k**, and Poynting vector **S** for a transverse electromagnetic (TEM) plane wave in an orthogonal system of vectors, as prescribed by Maxwell's equations. **a** Right-handed medium. **b** Left-handed medium

propagation constant) given by[7]

$$v_p = \frac{\omega}{\beta}, \tag{2.1}$$

and the group velocity (parallel to the direction of power flow or Poynting vector) defined as

$$v_g = \frac{d\omega}{d\beta}, \tag{2.2}$$

travel in opposite directions, a physical phenomenon that is called *backward-wave* propagation.

[7]In the literature the propagation constant is represented by k or β [8]. Hereafter the symbol β will be used since it is the usual convention in microwave engineering.

Therefore, LH media are characterized by the propagation of waves exhibiting anti-parallel[8] phase and group velocities, contrary to the case of wave propagation in conventional media that is known to be *forward* [1–4].[9] As a result, it follows that, since energy transfer is always outward from the source, the direction of propagation of the wavefronts in a LH medium is inward to the source. In consequence, the phase of a backward wave increases in the direction of group velocity (phase advance), in contrast to conventional materials which introduces a negative phase (phase lag). It is also worth mentioning that backward-wave propagation has implications in well known physical effects related to electromagnetic wave propagation, such as the reversal of Snell's law, Doppler effect, and Cherenkov radiation [1, 3].

The history of metamaterials may be considered to start in 1967 with the seminal work by Veselago [11], who examined the feasibility of media characterized by simultaneously negative permittivity and permeability. He concluded that such media are allowed by Maxwell's equations, establishing their fundamental properties and predicting some unusual electromagnetic phenomena, such as backward-wave propagation and left-handedness. Accordingly, he labeled conventional materials as RH media, and media with negative constitutive parameters as LH media.

Although Veselago may be considered the father of left-handed media [1], more than 3 decades elapsed until the first LH material was conceived and experimentally demonstrated by Smith et al. in 2000 [12]. This metamaterial was a composite structure of conducting wires and split-ring resonators (SRRs) [13], as shown in Fig. 2.3a. The effective permeability of an array of SRRs becomes negative in a narrow band above their resonance frequency, and a stopband response arises in the vicinity of that frequency. Conversely, an array of parallel thin metallic wires produces an effectively negative permittivity below a cutoff frequency,[10] exhibiting a highpass behavior. Thereby a proper combination of SRRs and wires (i.e. achieving a frequency region with simultaneously negative constitutive parameters), the stopband near the resonance of the SRRs switches to a passband with a LH signature (see Fig. 2.3b). Although this metamaterial is a bulk construction, since left-handed behavior occurs for a specific polarization and direction of propagation of the incident electromagnetic field, this structure can be viewed as a one-dimensional metamaterial. It should be noted that a LH medium made of natural media may also be possible. Unfortunately, the negative value of the constitutive parameters occurs at different frequency bands, preventing simultaneously negative permittivity and permeability [3].

---

[8] Anti-parallel vectors are collinear vectors (lying in the same line) with opposite directions.

[9] Backward-wave propagation has been known for decades. For instance, periodic structures support an infinite number of positive (forward) and negative (backward) space harmonics in addition to the fundamental space harmonic [7, 9]. The novelty of LH materials is that they are effectively homogeneous structures operating in a backward fundamental space harmonic [1, 2]. Forward/backward waves should not be confused with forward-/backward-traveling waves which are simply forward waves traveling in the positive/negative direction of propagation [10]. The latter are also known as positive- and negative-going waves [1].

[10] This cutoff frequency is usually called plasma frequency in analogy to the permittivity function of a plasma.

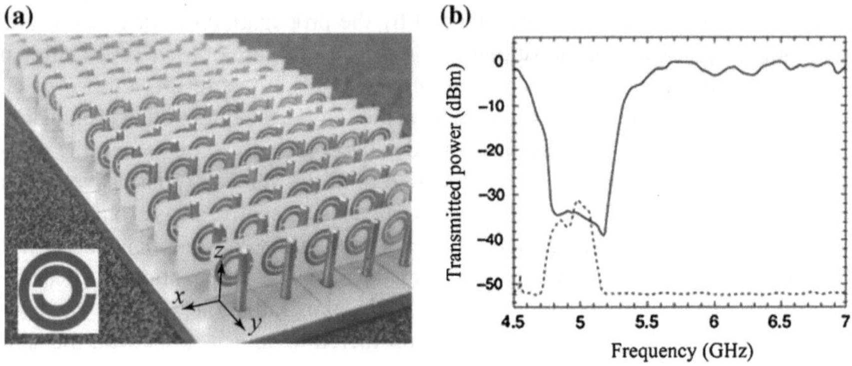

Fig. 2.3 First artificially synthesized LH structure. **a** Photograph reprinted with permission from [1]. **b** Transmitted power (x-axis orientation) when only SRRs are considered (*solid line*) and when wires and SRRs are present (*dashed line*). The incident electric field is parallel to the wires (z-axis), while the incident magnetic field is polarized along the axis of the rings (y-axis)

2.2 Transmission-Line Metamaterials

Transmission-line metamaterials (or *metamaterial transmission lines*)[11] are *artificial transmission lines* [6] that consist of a host (conventional) transmission line[12] loaded with reactive elements, the latter providing higher design flexibility in comparison to conventional lines [3, 4, 6]. Specifically, one-dimensional (1D) transmission-line metamaterials are effectively homogeneous structures whose electromagnetic characteristics can be controlled or engineered to some extent through a specific direction of propagation (this definition is, however, revised in Sect. 2.2.5).

2.2.1 Application of the Transmission-Line Theory to Metamaterials

Soon after the first experimental demonstration of left-handedness, several pioneering groups focused their research efforts on developing a transmission-line approach (non-resonant) of metamaterials exhibiting backward-wave propagation [14–16].

The reported approach originates from the fact that Maxwell's equations with plane-wave propagation in homogeneous and isotropic media have an identical form to the equations describing TEM propagation on a transmission line derived from

[11] In this thesis *transmission-line metamaterials* and *metamaterial transmission lines* are synonyms because this book deals with one-dimensional (1D) structures. When the same concept is extended to 2D and 3D dimensions, the structures are referred to as transmission-line metamaterials (i.e. metamaterial transmission lines apply only to 1D structures) [6].

[12] Note that host transmission lines are commonly implemented on an ordinary dielectric substrate whose permittivity is different from the effective permittivity of the metamaterial.

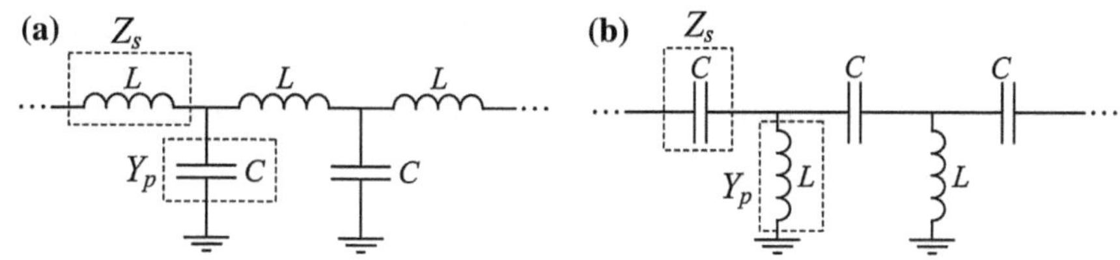

Fig. 2.4 Equivalent circuit model of a transmission line. **a** Conventional or RH transmission line. **b** Dual or LH transmission line

circuit theory, known as the telegrapher equations. Hence, such an analogy allows the series and shunt elements of the well-known ladder circuit model of a transmission line to be related to the constitutive parameters of a medium (exhibiting the same propagation characteristics) by mapping the telegrapher equations to Maxwell's equations. The resulting relationships are [1–4]

$$Z_s' = Z_s/l = j\omega\mu, \tag{2.3a}$$
$$Y_p' = Y_p/l = j\omega\epsilon, \tag{2.3b}$$

where Z_s' and Y_p' are the (distributed) per-unit-length series impedance (Ω/m) and shunt admittance (S/m), respectively, Z_s and Y_p are the per-unit-cell series impedance (Ω) and shunt admittance (S), respectively, and l is the unit cell length (the period).

Accordingly, in conventional RH media, the mapping yields

$$\mu = L' = L/l, \tag{2.4a}$$
$$\epsilon = C' = C/l, \tag{2.4b}$$

where L' and C' are the per-unit-length series inductance (H/m) and shunt capacitance (F/m), respectively, while L and C represent the per-unit-cell series inductance (H) and shunt capacitance (F), respectively. The resulting well-known equivalent circuit is shown schematically in Fig. 2.4a, whose propagation is forward with a propagation constant and a characteristic impedance given by

$$\beta_R = \omega\sqrt{L'C'}, \tag{2.5a}$$
$$Z_{cR} = \sqrt{\frac{L'}{C'}}, \tag{2.5b}$$

respectively. In the transmission-line approach of metamaterials the previous well-established analogy has been extended to the other (exotic) combinations of the constitutive parameters depending on their sign. Thus, for a LH medium, the series inductance and the shunt capacitance should become negative according to (2.4). Since from an impedance perspective a negative inductance/capacitance may be interpreted as a positive capacitance/inductance [2], a ladder network as the one depicted in Fig. 2.4b with series-connected capacitors and shunt-connected inductors supports the propagation of backward waves. Such a network is called the *dual*[13] of the equivalent circuit of a RH transmission line. The mapping in this dual network yields

$$\mu = -\frac{1}{\omega^2 C'} = -\frac{1}{\omega^2 Cl}, \tag{2.6a}$$

$$\epsilon = -\frac{1}{\omega^2 L'} = -\frac{1}{\omega^2 Ll}, \tag{2.6b}$$

where C' and L' are now the time-unit-length[14] series capacitance ($\text{F} \cdot \text{m}$) and shunt inductance ($\text{H} \cdot \text{m}$), respectively, while C and L stand for the per-unit-cell series capacitance (F) and shunt inductance (H), respectively. The corresponding propagation constant and characteristic impedance become

$$\beta_L = -1/\omega\sqrt{L'C'}, \tag{2.7a}$$

$$Z_{cL} = \sqrt{\frac{L'}{C'}}, \tag{2.7b}$$

respectively. Contrarily to a dispersionless RH line, a LH transmission line is intrinsically dispersive by nature because its propagation constant is not a linear function of frequency. It should also be emphasized that the mapping given in (2.3) is valid only in the long wavelength regime, where the period is much smaller than the guided wavelength

$$\lambda_g = \frac{2\pi}{|\beta|}, \tag{2.8}$$

which is defined only in the passbands [1]. Only under this circumstance the definition of effective constitutive parameters is fully feasible and a ladder circuit may describe a RH or LH transmission line.

Figure 2.5 summarizes the generalization [2] of the aforementioned transmission-line approach of metamaterials by means of unit-cell equivalent T-circuit models. Besides the RH (or DPS) and LH (or DNG) transmission lines, the equivalent circuits of SNG transmission lines are also indicated. It should be noted that waves cannot propagate in either ENG or MNG transmission lines because their models are exclusively made up of inductors or capacitors. As wavelength cannot be defined, these models should be interpreted conceptually. To sum up, the transmission-line theory in metamaterials links field to circuit quantities, offering an alternative interesting physical interpretation leading to the same functional results.

---

[13]Duality is defined here in terms of complementary response (e.g. the dual of a lowpass filter is a highpass filter) so that the series/shunt impedance of a network is proportional to the series/shunt admittance of the other network.

[14]This ensures properly defined units in Z_s' (Ω/m) and Y_p' (S/m).

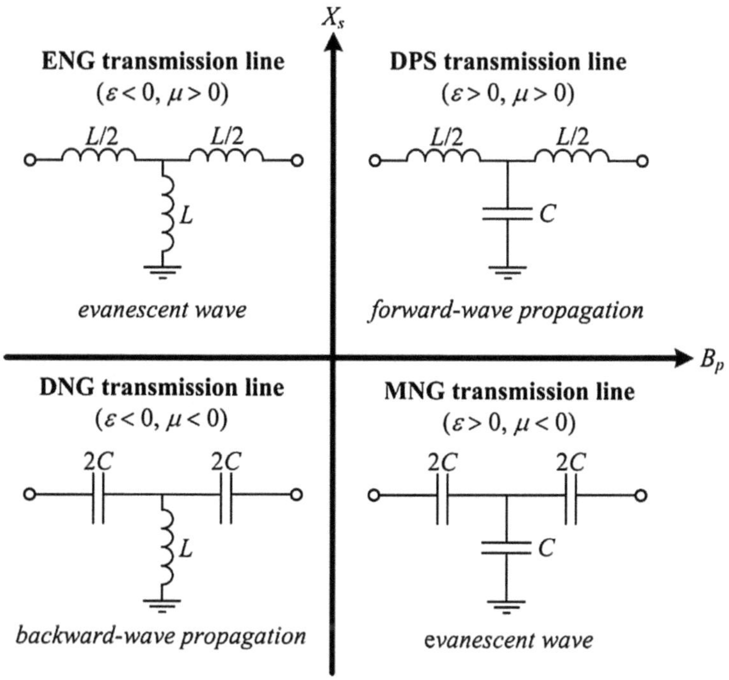

Fig. 2.5 Unit-cell equivalent T-circuit models of transmission lines according to the sign of the series reactance, X_s ($Z_s = jX_s$), and shunt susceptance, B_p, ($Y_p = jB_p$). The validity of these models is restricted to the long wavelength regime. Since wavelength cannot be defined if propagation is forbidden, as occurs in ENG and MNG transmission lines, these models are conceptual interpretations

2.2.2 Composite Right-/Left-Handed (CRLH) Transmission Lines

From a practical point of view, a purely left-handed (PLH) transmission line as that depicted in Fig. 2.4b cannot be implemented. A necessary host transmission line introduces inevitably parasitic elements, as illustrated in Fig. 2.6 [1–4]. Since these elements cannot be neglected in general and induce wave propagation of forward nature, practical realizations of left-handed transmission lines exhibit left- or right-handed propagation, depending on the frequency interval. For this reason, such lines are termed composite right-/left-handed (CRLH) transmission lines [17], and they constitute the most general transmission-line metamaterial [1]. Indeed, as will be shown shortly, a CRLH line is able to behave as a DPS, DNG, or ENG/MNG transmission line within particular frequency ranges.

Instead of describing the transmission characteristics of CRLH lines obtaining their effective constitutive parameters by (2.3), Fig. 2.7 shows the typical dispersion

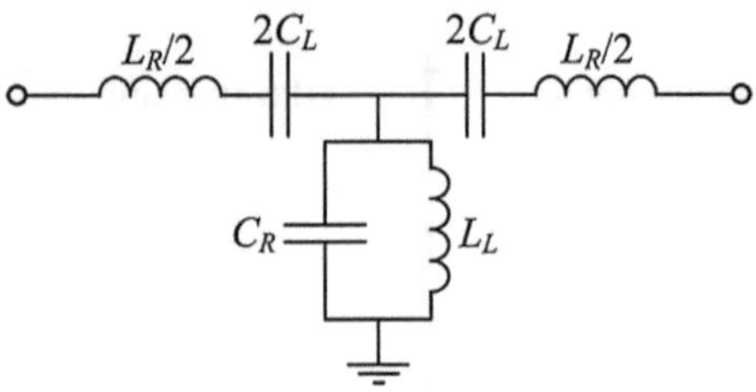

Fig. 2.6 Unit-cell equivalent T-circuit of a composite right-/left-handed (CRLH) transmission line; the subscripts L and R stand for the corresponding LH and RH elements, respectively

diagram[15] of a CRLH line, which indicates the variation of the propagation constant with frequency. Hence, the dispersion diagram reveals the passbands/stopbands as well as the nature (RH or LH) of the passbands since the phase and group velocities can be easily inferred from (2.1) and (2.2), respectively.[16] As can be seen, if $\omega < \min(\omega_s, \omega_p)$, $v_p < 0$ and $v_g > 0$ have opposite signs, meaning that propagation is backward. Conversely, when $\omega > \max(\omega_s, \omega_p)$, $v_p > 0$ and $v_g > 0$ exhibit the

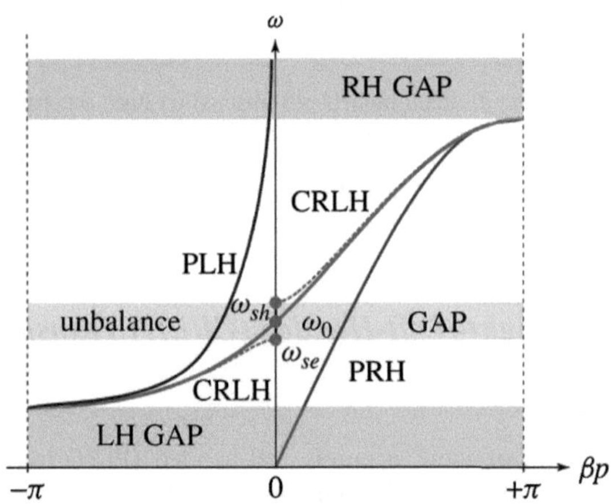

Fig. 2.7 Typical dispersion diagram for a unbalanced/balanced CRLH transmission line. The period is $p = l$, while the series and shunt resonances are indicated as $\omega_{se} = \omega_s = 1/\sqrt{L_R C_L}$ and $\omega_{sh} = \omega_p = 1/\sqrt{L_L C_R}$, respectively. The case with $\omega_s < \omega_p$ is illustrated, although $\omega_s > \omega_p$ is also possible. The curves for a purely RH (PRH) line ($L_L = C_L \rightarrow \infty$) and for a purely LH (PLH) line ($L_R = C_R = 0$) described by (2.5a) and (2.7a), respectively, are also shown. Reprinted with permission from [2]

---

[15] The calculation procedure of the dispersion relation of a periodic structure composed of a cascade of two-port unit-cell circuits is well-known (Appendix A).

[16] The phase velocity is the slope of the line segment from the origin of the dispersion curve $\omega(\beta)$ to a point in the curve, whereas the group velocity is the slope of the tangent to the dispersion curve at a point.

same sign, indicating that wave propagation is of forward nature.[17] The frequency at the transition from a passband to a stopband ($\beta l = 0$ or $\pm\pi$) is often called cutoff frequency [10], being the group velocity equal to zero. There are some stopbands (or bandgaps) in the CRLH line, characterized by evanescent waves caused by reflections. When $\omega \to 0$ there is a stopband due to the highpass nature of the LH elements, whereas when $\omega \to \infty$ a stopband arises because of the lowpass behavior of the RH components. It is thus apparent that the transmission characteristics of a CRLH line at low and high frequencies tend to behave as purely LH and RH transmission lines, respectively (this is due to periodicity, since a RH transmission line exhibits an all-pass response at all frequencies). In addition, in the general case when the series resonance, ω_s, does not coincide with the shunt resonance, ω_p, a stopband appears between these frequencies, and the line is known to be *unbalanced.* By contrast, in the case that these frequencies are equal, $\omega_s = \omega_p = \omega_0$, there is a continuous transition between the left- and right-handed bands,[18] and the CRLH line is said to be *balanced*[19] [1–3].

The guided wavelength,[20] given by (2.8), can also be readily derived from the dispersion curve. As mentioned in Sect. 2.1, an effectively homogeneous structure is defined as that whose unit cell size l is much smaller than λ_g. The effective-homogeneity condition is usually $l < \lambda_g/4$ [1], which in terms of the electrical length becomes $|\beta l| < \pi/2$. Since in the unbalanced configuration the structure is effectively homogeneous at the edges of the bandgap that ranges from ω_s to ω_p, this stopband in terms of the constitutive parameters is said to be due to either $\epsilon < 0$ (if $\omega_s < \omega_p$) or $\mu < 0$ ($\omega_s > \omega_p$) [1]. By contrast, the stopbands at low and high frequencies are not near the long wavelength regime. Therefore, at these frequencies, effective constitutive parameters cannot be rigorously defined and the circuit model of Fig. 2.6 is no longer valid to represent an effectively homogeneous CRLH transmission-line metamaterial. In order to model correctly RH/LH (or CRLH) transmission lines at higher/lower frequencies, one needs to reduce the period, and accordingly to reduce/increase the per-section inductance and capacitance given in (2.4)/(2.6), so that operation in the long wavelength regime holds. In other words, the circuit models of RH/LH/CRLH transmission lines mimic transmission lines only in their passbands where homogeneity is satisfied. Note that the circuit model of a CRLH transmission line is the same as that of a bandpass filter, and the circuits of RH and LH lines are lowpass and highpass filters, respectively [18]. Classical filters, however, are characterized at all frequencies (from $\omega = 0$ to $\omega \to \infty$). Therefore, as a fundamental property, filters exhibit both passband and stopband characteristics according to the filter type.

[17] Alternatively, when $|\beta|$ increases with frequency, v_p and v_g are parallel and the propagation is forward. Otherwise the propagation is backward.

[18] Wave propagation is allowed at ω_0 with infinite v_p but (non-zero) finite v_g.

[19] The term *balanced* does not mean *differential* in this context.

[20] The guided wavelength is obviously different from that guided in the host line ($\beta \neq \beta_R$).

2.2.3 CL-Loaded and Resonant-Type Approaches

Because CRLH transmission lines are not readily available in nature, these structures have to be implemented by means of a host transmission line loaded with reactive elements (inductors, capacitors, and/or both of them in the form of resonators). Experimental CRLH lines were reported for the first time by Caloz et al. in microstrip technology [16], and by the Group of Eleftheriades in coplanar waveguide implementation [19]. The reactive elements responsible for backward propagation in both configurations were synthesized by semi-lumped[21] components, such as series capacitive gaps and shunt metallic inductive strips (discrete commercial chip components can also be used instead [1]). This direct implementation is referred to as *LC network implementation* [1] or *CL-loaded approach* [3, 6]. Alternatively, transmission-line metamaterials featuring a CRLH behavior can also be built up by the so-called *resonant-type approach* [3, 6]. This realization consists of a host transmission line loaded with subwavelength resonators[22] in combination with other reactive elements (inductors or capacitors), the typical resonant particles being split-ring resonators (SRRs) or other related topologies as those described in Sect. 2.3.[23] The first reported structure concerning this approach was proposed by Martín et al. [20], and it was conceptually based on and equivalent to the preceding Smith's LH structure shown in Fig. 2.3. Specifically, a coplanar waveguide (CPW) was periodically loaded with pairs of SRRs and signal-to-ground inductive strips (see Fig. 2.8). The SRRs were etched on the back side of the substrate, and underneath the slots of the CPW in order to enhance line-to-resonator magnetic coupling (this is the main coupling mechanism of SRRs as will be reviewed in Sect. 2.3.1 [21]). Another possibility to implement a CRLH response based on the resonant-type approach is by means of series capacitive gaps, and the complementary version (see Sect. 2.3.5) of the SRR, that is, the so-called complementary split-ring resonator (CSRR) [22, 23]. A chain of CSRRs provides a negative effective ϵ, while a negative effective μ is achieved through the gaps. Such a kind of line is illustrated in Fig. 2.9, where the CSRRs are etched in the ground plane of a microstrip line, beneath the signal strip, as strong line-to-resonator electric coupling (the dominant excitation field of CSRRs [21]) is achieved.

---

[21] Semi-lumped components are planar elements whose physical dimensions are usually restricted to be smaller than a quarter of wavelength. The components may also be designated to as lumped elements when the dimensions are smaller than one-eighth wavelength. These terminologies are in contrast to distributed (transmission line) elements in which the physical size is comparable to or larger than the wavelength [10, 18]. The advantage of electrically small planar (i.e. semi-lumped or even lumped) elements is that their behavior may be approximate by ideal (sizeless) lumped parameters using circuit theory.

[22] Subwavelength resonators are those whose physical size is a small portion of the wavelength, and accordingly they may be referred to as semi-lumped or lumped resonators when composed of semi-lumped or lumped inductors and capacitors, respectively [18].

[23] Note that CRLH lines based on the *CL*-loaded approach are also resonant by nature despite the fact that self-resonant elements are not used.

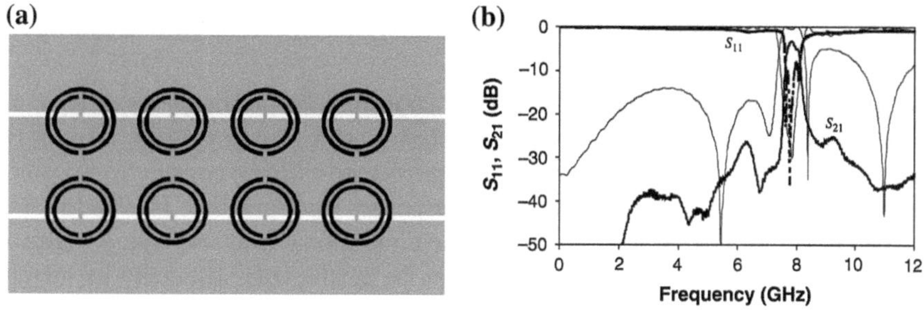

Fig. 2.8 CRLH transmission line based on SRRs and shunt inductive strips in coplanar waveguide technology. **a** Layout. **b** Simulated (*thin line*) and measured (*thick line*) frequency response. The passband is LH, since the CRLH line is not balanced and the RH passband is beyond the shown frequency range. Reprinted with permission from [20]

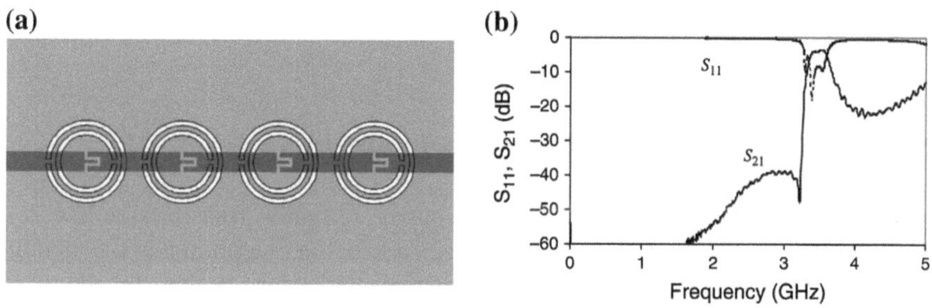

Fig. 2.9 CRLH transmission line based on CSRRs and series capacitive gaps in microstrip technology. **a** Layout reprinted with permission from [21]. **b** Measured frequency response reprinted with permission from [3]. The RH passband is outside the measured frequency range

Since the resonant-type approach employs subwavelength resonators, the individual unit cells can be modeled by lumped-element equivalent circuit models. However, the frequency response of a CRLH transmission line obtained by the resonant-type approach does not obey exactly the same behavior as the canonical[24] circuit of Fig. 2.6. Nevertheless, the corresponding models are qualitatively quite similar to the canonical one. The meaningful difference is the presence of a transmission zero on the left-hand side of the backward-wave band [3], which naturally is very useful in filtering applications. The first equivalent circuit models of SRR- and CSRR-based CRLH transmission lines were reported by Martín et al. [20] and Baena et al. [21], respectively, and these models were later revised and improved by Aznar et al. [24] and Bonache et al. [25], respectively. The main improvement is relative to the fact that by adding shunt strips (in SRR-loaded CPWs) and series gaps (in CSRR-loaded microstrip lines), the transmission zero is displaced, and this fact was not predicted by the initial models. None of these circuits models is shown here, but they are very similar to those of SNG transmission lines that will be presented in the next subsection.

[24]Canonical refers to the simplest or standard form.

2.2.4 Resonant-Type Single-Negative Transmission Lines

A CRLH structure is intended to be used as a transmission line in the sense that only the passband/s is/are directly useful, the stopbands being usually parasitic effects [1]. Instead, structures exhibiting equivalent single negative (SNG) constitutive parameters are preferred to implement stopband responses (the structures involved in this thesis belong, in fact, to this category). One-dimensional resonant-type μ-negative (MNG) and ϵ-negative (ENG) structures can be synthesized artificially by merely loading a host line with SRRs and CSRRs, respectively. Signal propagation in these structures is inhibited in a narrow band in the vicinity of the resonance frequency of the resonators, whereas wave propagation is forward outside the forbidden band. Such a narrowband response is of course due to the self-resonant behavior of the loading particles. Consequently, these structures behave qualitatively as the canonical circuit models describing SNG transmission lines depicted in Fig. 2.5 only within a certain frequency range.

2.2.4.1 μ-Negative Coplanar Waveguides Loaded with Pairs of SRRs

Figure 2.10a shows the first proposed SRR-based μ-negative transmission line. It is the same structure as the CRLH line in Fig. 2.8a with the exception that the metallic strips are not introduced. Clearly the most relevant feature is that, by removing such strips, the backward passband in the vicinity of the SRR resonance frequency switches to a stopband response, as shown in Fig. 2.10b [20, 26].

In order to gain more insight into the transmission and reflection properties of SRR-loaded lines, the equivalent series reactance and shunt susceptance of a typical unit cell are plotted in Fig. 2.11a. The dispersion relation corresponding to an infinite structure obtained by cascading identical unit cells is shown in Fig. 2.11b. As can be observed, there is a stopband in the vicinity of the resonance frequency, f_0, which occurs when the length of the unloaded line is a small fraction of the guided

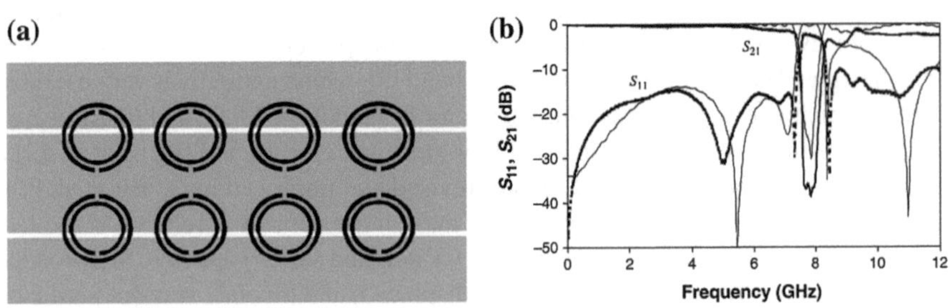

Fig. 2.10 MNG transmission line based on SRRs in coplanar waveguide implementation. **a** Layout. **b** Simulated (*thin line*) and measured (*thick line*) frequency response. Reprinted with permission from [26]

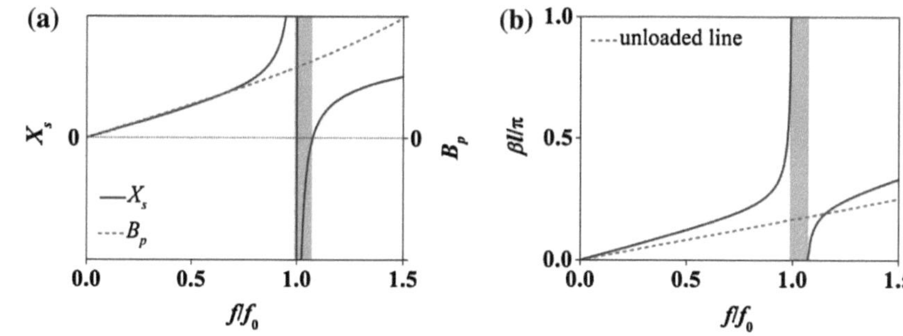

Fig. 2.11 Typical behavior of a unit-cell SRR-loaded transmission line. **a** Series reactance, X_s, and shunt susceptance, B_p, of an equivalent π-network extracted from electromagnetic simulation. **b** Dispersion diagram assuming an infinite periodic structure; the dispersion of the unloaded transmission line section is also shown. The stopband is highlighted in *gray*

wavelength (the dispersion for the considered transmission line section without the SRR is also shown). An analysis in terms of the impedances is as follows. Beyond f_0, the series impedance is capacitive in a narrow band, and signal propagation is inhibited because the shunt impedance is also capacitive. However, the stopband region extends also below f_0, despite the fact that the series and shunt impedances behave inductively and capacitively, respectively; the reason is that the series impedance attains extreme inductive values.[25] In relation to the constitutive parameters, the stopband has been associated to a negative μ in a narrow region above f_0 [21]. Within this frequency band the structure behaves as the canonical MNG transmission line shown in Fig. 2.5, since the structure is effectively homogeneous near this band ($\beta l = 0$ at the lower edge of the second propagating band). It is on occasions said that the stopband is also due to an extremely high positive μ in a narrow frequency range below f_0. However, this statement is somewhat misleading in a strict sense, since as the series impedance becomes more and more inductive the operation regime goes away from the homogeneity condition ($\beta l \rightarrow \pi$).

Due to the small electrical dimensions of the SRRs at resonance, the behavior of SRR-loaded lines can be explained by the unit-cell equivalent-circuit model depicted in Fig. 2.12a. This circuit model was initially proposed by Martín et al. [20, 26], and later revised by Aznar et al. [24]. The per-section inductance and capacitance of the line are modeled by L and C, respectively, and the SRR is described by the resonant tank $L_s - C_s$ (see Sect. 2.3.1) that is magnetically coupled to the line through the mutual inductance M. This circuit can be simplified to the equivalent one depicted

[25]Simultaneously positive or negative values of ϵ and μ are necessary and sufficient for wave propagation, as was illustrated in Fig. 2.1 (operation in the long wavelength regime is implicitly fulfilled). By contrast, the same sign of X_s and B_p (or different sign of X_s and $X_p = -1/B_p$) cannot guarantee wave propagation, unless the operation band is explicitly restricted to the long wavelength regime, as highlighted in Fig. 2.5. Note that possible impedance mismatch with feeding sources at the port terminals in practical situations resulting in reflections is not considered in the aforementioned assertions.

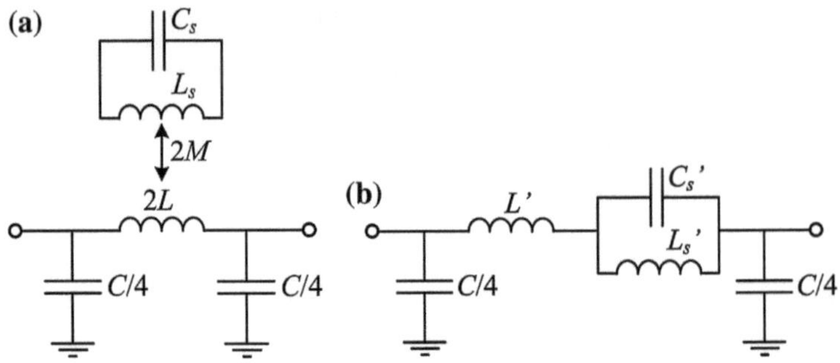

Fig. 2.12 Unit-cell equivalent-circuit model of an SRR-loaded CPW. **a** Model corresponding to half of the structure since a *magnetic wall* (open circuit) is applied along the longitudinally-oriented midplane by virtue of the symmetric topology and excitation mode (an even mode). **b** Equivalent transformed model using (2.9)

in Fig. 2.12b, provided that [3, 24, 27]

$$L' = 2L - L'_s, \tag{2.9a}$$

$$L'_s = 4C_s(\omega_0 M)^2 = 4\frac{M^2}{L_s}, \tag{2.9b}$$

$$C'_s = \frac{L_s}{4(\omega_0 M)^2} = \frac{C_s}{4}\left(\frac{L_s}{M}\right)^2, \tag{2.9c}$$

where

$$\omega_0 = \frac{1}{\sqrt{L_s C_s}} = \frac{1}{\sqrt{L'_s C'_s}} = 2\pi f_0, \tag{2.10}$$

is the angular resonance frequency of the SRR. Note that a zero in the transmission coefficient ($S_{21} = 0$) arises at this frequency, since the series branch is open-circuited. Therefore, the transformed circuit model clearly explains that the excitation of the SRR is responsible for returning the injected power to the source. Moreover, the lumped-element values of this circuit can be easily extracted systematically [28] (this procedure is used throughout this thesis). Finally, it should be noted that the circuit model converges to that of MNG transmission lines within a narrow bandwidth above resonance.

2.2.4.2 ϵ-Negative CSRR-Loaded Microstrip Lines

The first reported CSRR-loaded microstrip line is shown in Fig. 2.13a, which is the same as the one shown in Fig. 2.9a but with a uniform conductor strip (i.e. without the gaps). As can be seen in Fig. 2.13b, by removing the series gaps, signal propagation is inhibited in a narrow frequency band in the vicinity of the CSRR resonance frequency.

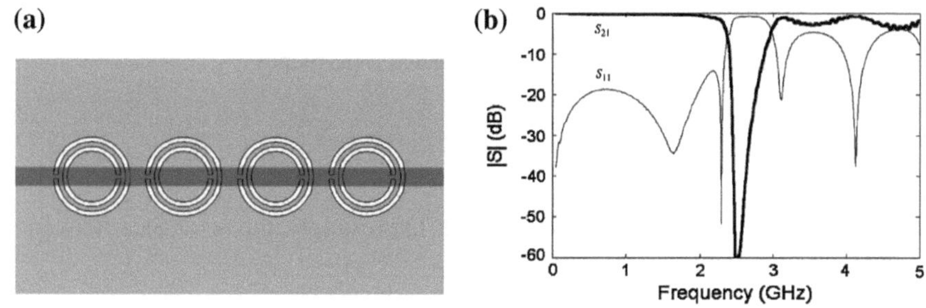

Fig. 2.13 ENG transmission line based on a CSRR-loaded microstrip line. **a** Layout and **b** measured frequency response. Reprinted with permission from [22]

In order to inquire into the propagation characteristics of CSRR-based transmission lines, the typical series reactance, shunt susceptance, and dispersion relation are plotted in Fig. 2.14. As can be observed, the stopband in the vicinity of the transmission zero frequency, f_z, is due to a shunt susceptance with negative ($f > f_z$) and extremely positive ($f < f_z$) values. Regarding the constitutive parameters, a CSRR-loaded line has been interpreted to behave in the stopband ($f > f_z$) as a 1D ϵ-negative metamaterial (where the structure may be viewed as the canonical ENG transmission line schematically depicted in Fig. 2.5) [21].

As long as the CSRRs are electrically small at their fundamental resonance, CSRR-loaded lines can be modeled by the lumped-element unit-cell circuit shown in Fig. 2.15 [21]. The CSRR is represented by the resonant tank $L_c - C_c$ (Sect. 2.3.5), and the per-section line inductance and capacitance are modeled by L and C, respectively. More specifically, C in reality accounts for the line-to-resonator electric coupling as this coupling can be modeled by connecting in series the line capacitance to the CSRR (see Appendix B.1). The extraction of the values of the circuit parameters

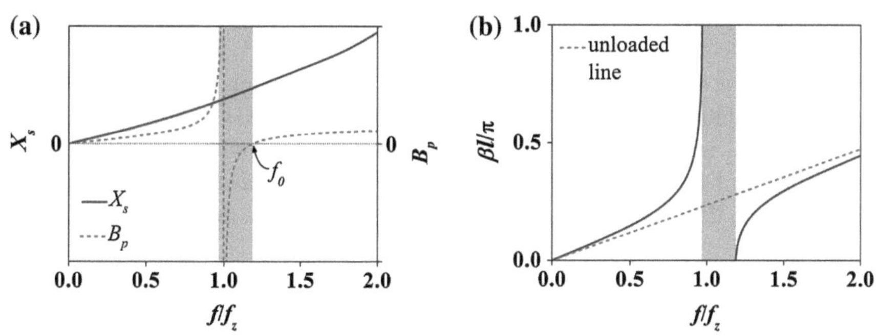

Fig. 2.14 Typical behavior of a unit-cell CSRR-loaded transmission line. **a** Series reactance, X_s, and shunt susceptance, B_p, of an equivalent T-network extracted from electromagnetic simulation. **b** Dispersion diagram for an infinite periodic structure; the dispersion of the unloaded transmission line section is also shown. The stopband is highlighted in *gray*

for a particular CSRR-loaded line is straightforward [29]. Inspection of this circuit model reveals that there is a transmission zero at a frequency

$$\omega_z = \frac{1}{\sqrt{L_c(C + C_c)}} = 2\pi f_z, \tag{2.11}$$

at which the shunt branch is short-circuited. Interestingly, the resonance frequency of the CSRR, given by

$$\omega_0 = \frac{1}{\sqrt{L_c C_c}} = 2\pi f_0, \tag{2.12}$$

occurs at a higher frequency than the transmission zero ($f_0 > f_z$).

Section 2.2.4 has been focused on SRR-loaded CPWs and CSRR-loaded microstrip lines. In these configurations, a significant component of the dominant excitation field (namely, magnetic in SRRs and electric in CSRRs) results parallel to the resonator axis.[26] Nonetheless, it is important to highlight that the corresponding circuit models and parameter extraction procedures are *general*. This means the models may be valid regardless of the resonator topology (Sect. 2.3) and the host line technology (e.g. CPW or microstrip) if the resonator is electrically small and coupled either magnetically (as the SRR) or electrically (as the CSRR) to the line. With regard to the circuit parameter extraction methodologies, despite the fact that they require the electromagnetic simulation of the structures, they are versatile in the sense that there is no restriction in the structures involved, in contrast to what usually occurs in analytical models. Therefore, the transmission and reflection characteristics of SRR- and CSRR-loaded lines are indeed somewhat general and may be extrapolated to other related structures.

In the light of Figs. 2.11a and 2.14a, it should be stressed that the proposed circuit models are able to mimic the behavior of SRR- and CSRR-loaded lines up to beyond the resonance frequency and the transmission zero frequency (the region of interest). In other words, the validity of the circuits is not limited to the long wavelength region.[27] To end this section, it is interesting to realize that the circuit models of SRR- and CSRR-loaded lines (Figs. 2.12 and 2.15) are formally dual circuits, that is, the series/shunt impedance of one of these models is proportional to the shunt/series admittance of the other circuit (compare Figs. 2.11a and 2.14a).[28] Hence, although SRR- and CSRR-loaded lines are not structurally complementary,[29] their corresponding behaviors manifest the duality property because the complementary resonators are illuminated by dual fields (this supports that the equivalent circuits

[26]The electromagnetic field distribution of these lines is analyzed in Sect. 4.1.

[27]The valid frequency range of the models is evaluated throughout the thesis by comparing circuit simulations with full-wave electromagnetic simulations and measurements.

[28]Two networks are (formally and numerically) dual if $Z_{ii} = Y_{ii}$ and $Z_{ij} = -Y_{ij} (i \neq j)$, where Z_{ij} are the impedance matrix parameters of one network and Y_{ij} are the admittance matrix parameters of its dual one [18].

[29]Besides the fact that complementary structures are defined in single planes, coplanar strips (CPS) are the complementary counterpart of a CPW.

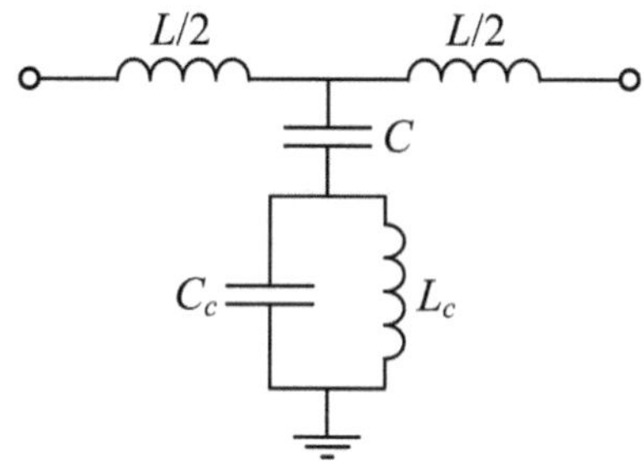

Fig. 2.15 Unit-cell equivalent-circuit model of a CSRR-loaded microstrip line

may be independent of the technology of the host transmission line). Thus, the dispersion relation of SRR- and CSRR-loaded lines is the same (at least qualitatively as is apparent from Figs. 2.11b and 2.14b), a property derived from dual circuits. When formally dual circuits are also numerically exact dual, their dispersion relation is exactly the same (Appendix A).

2.2.5 Discussion About Homogeneity and Periodicity

Homogenization in metamaterials means that the homogeneity condition is achieved in the vicinity of the operating frequency. As has been shown, the transmission-line approach of metamaterials characterizes the electromagnetic properties of homogeneous transmission-line metamaterials by either effective constitutive parameters or immittances.[30] However, this description is limited to the long wavelength regime, and cannot be extended to frequencies where the wavelength becomes comparable to the period. Obviously, such a restriction may mask and forbid useful functionalities, since some interesting phenomena occur outside the long wavelength regime. Contrarily, the propagation constant as well as the characteristic impedance of periodic structures (including transmission-line metamaterials) can be defined rigorously at all frequencies using the powerful classical theory of periodic structures [30] (Appendix A), without invoking the theory of effective-medium approximation. Therefore, from an engineering standpoint, periodic metamaterial structures are usually characterized by their propagation constant and characteristic impedance rather than by their effective permittivity and permeability [1, 3, 4].

In general, the synthesis of LH (or CRLH) devices is usually referred to as *dispersion and impedance engineering* [3, 4, 6] since the relevant parameters are the electrical length and the characteristic impedance. Indeed, for the design of circuits and components, homogeneity is irrelevant to tailor the electrical length and the impedance. Furthermore, these devices are commonly composed of a single unit

[30] Immittances are either impedances or admittances [18].

cell[31] toward miniaturization, leading to indisputably inhomogeneous structures. With respect to stopband structures (SNG-based), by sacrificing homogeneity, significant signal rejection may be achieved within a determined forbidden band by a few periods, making unnecessary a large structure composed of many cells.

To sum up, the design of devices in the RF/microwave community is not usually focused on the implementation of effectively homogeneous transmission lines, but on taking advantage of their high controllability in the response, and/or their small electrical size. When the effective-homogeneity condition is not satisfied and/or there are not many cells, these devices cannot be strictly considered as metamaterials. However, as suggested by Martín et al. [6] and Aznar et al. [31], the definition of metamaterial transmission lines is not usually restricted to those exhibiting homogeneity and periodicity or quasi-periodicity (contrary to the definition given at the beginning of Sect. 2.2). Consequently, terms such as *transmission-line metamaterials*, *metamaterial transmission lines*, *metamaterial-based lines*, *metamaterial-inspired lines*, or even *metalines*, are often used in a broad sense to describe artificial transmission lines based on metamaterial concepts, ignoring whether effective homogeneity and periodicity are accomplished or not.[32]

2.3 Metamaterial-Based Resonators

This section analyzes several resonator topologies in the form of split rings[33] employed for the synthesis of metamaterials, transmission-line metamaterials, or RF/microwave devices inspired by metamaterials. These resonators are often known as *metamaterial-based resonators* or simply *metamaterial resonators*. Indeed, negative permeability/permittivity structures may be implemented by means of topologies different from that of the original SRR[34]/CSRR. The additional considered resonant elements are *strip* (made of an arrangement of metals) as well as *slot* (engineered apertures in a metallic plane) resonators. Many electrically small planar resonators have been reported in the literature [3, 5, 6], and they can be classified according to several criteria. The list is limited here to those split ring type resonators that are used throughout this thesis. All of them have in common the following features:

- *Subwavelength*, and accordingly they may be referred to as semi-lumped or lumped resonators.
- *Closed*, in the sense that resonance is induced by external electromagnetic fields.[35]

---

[31]The propagation constant and the characteristic impedance obtained from the theory of periodic structures is still valid for a single unit cell.

[32]Besides the broad definition, a distinction between metamaterial-based and metamaterial-inspired has been proposed [32].

[33]*Split rings* are also sometimes called *open rings* [33].

[34]*SRR* is exclusively used here to designate the original split-ring resonator topology [13].

[35]On the contrary, *open* split-ring resonators have well-defined port terminals, e.g. the open split-ring resonator (OSRR) [34] or the open complementary split-ring resonator (OCSRR) [35].

- *Planar*, being fully compatible with planar technology.
- *Symmetric*, so that when the topology is bisected along its symmetry plane, one of the halves is the mirror image of the other half (a plane of symmetry acts as a mirror). Additionally, resonators having two symmetry planes are said to be *bisymmetric*.

However, all the topologies exhibit different peculiarities that make them useful or more appropriate in some circumstances. The main purpose of this section is to gain insight into the electromagnetic characteristics and the equivalent circuit models of the resonators derived from symmetry properties (a discussion about different alternatives for the synthesis of metamaterial structures is not provided). With a view to doing so, the SRR topology still holds special attention because of its relevance.

The most basic assumption in effective media theory (Sect. 2.1) is that homogenization (averaging) makes sense only if the variation of the average field is small at the scale of the period [3, 13]. The validity of homogenization is usually expressed in terms of the size of the elements forming the discrete medium, so that they must be small as compared with the wavelength. It is assumed here that the size of the resonators is on a scale much shorter than the wavelength, such that an average value for the fields may be sensibly defined. Thus, the resonators are supposed to be illuminated by external uniform (i.e. spatially constant in magnitude and phase) time-varying electric or magnetic fields with the customary sinusoidal, or harmonic, time dependence $e^{j\omega t}$.

2.3.1 Split-Ring Resonator (SRR)

The split-ring resonator (SRR) consists of two metallic concentric and coupled split rings with slits on opposite sides, as depicted in Fig. 2.16a. Although there are some precedents in the scientific literature related to split-ring resonators [5, 36], this specific topology was proposed by Pendry et al. [13] to build up for the first time a negative permeability medium [12].

When the rings are excited by an external time-varying magnetic field directed along the axial direction (z-oriented), an electromotive force around the rings is generated that, in turn, induces currents in the rings. The slit (or gap) in each ring prevents current from flowing around the individual rings, so that the slits force the electric current to flow from one ring to the other through the inter-ring slot, taking the form of a displacement current (see Fig. 2.16b). As is apparent from Figs. 2.16b and 2.17, the symmetry plane (xz-oriented) behaves as a virtual *electric wall*. The current loops induce a magnetic dipole moment along the resonator axis (z-orientation). Additionally, an electric dipole moment is also generated in the plane of the particle, oriented orthogonally to the plane of symmetry (y-direction).[36]

[36]The radiation of split-ring resonators can be quantified by the induced dipole moments through an equivalent radiation resistance. Nevertheless, since the dimensions of the dipoles are small relative to the wavelength, in general radiation may be neglected [5, 37].

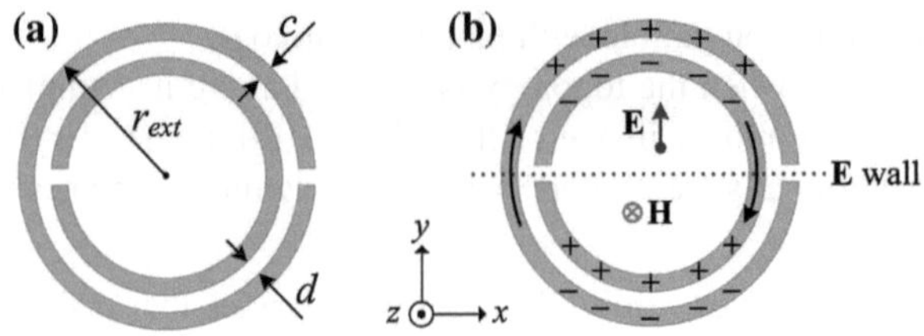

Fig. 2.16 Symmetric split-ring resonator (SRR). **a** Circular topology. **b** Fundamental resonance: orientation of the polarization fields, sketch of the charge distribution and currents, and boundary condition at the symmetry plane

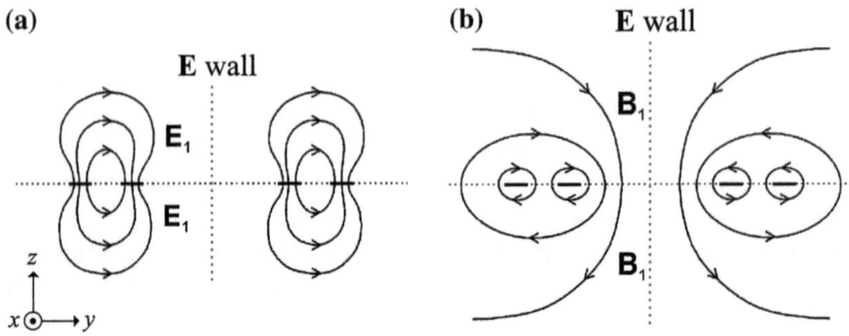

Fig. 2.17 Sketch of the electric **a** and magnetic **b** field distributions at the SRR fundamental resonance frequency. Reprinted with permission from [21]

By symmetry, the electric charges in the upper half of the SRR are the images of the charges in its lower half, and two parallel electric dipoles are generated on each half. This means that a time-varying electric field is also able to excite the SRR, inducing a charge density in the form of an electric dipole that causes currents to flow. When there is present an/a electric/magnetic dipole in a metamaterial resonator, the particle is said to be electrically/magnetically polarized.[37] Since both an electric dipole and a magnetic dipole are induced, the SRR suffers from cross-polarization effects, so that it may become electrically polarized as a response to an applied magnetic field, and vice versa [5, 38]. This property may also be derived from the fact that the SRR is not invariant by inversion, since any resonator without inversion symmetry with regard to its center must exhibit some degree of cross-polarizability [3].[38] However, although the SRR shows simultaneously magnetoelectric response at resonance, the magnetic field has been found to be the dominant excitation mechanism [21, 39, 40].

[37] Polarization is related here to the generation of local (microscopic) dipole moments in the presence of external fields (like medium electric and magnetic polarizations accounted for by ϵ and μ, respectively, at the macroscopic level), and not to wave polarization.

[38] A bidimensional object possesses inversion symmetry when is invariant by rotating 180° taking its center as the rotation axis.

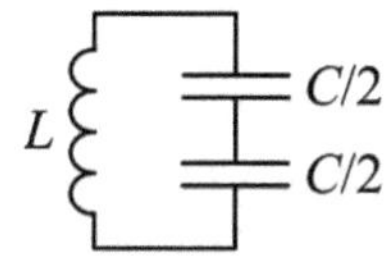

Fig. 2.18 Equivalent circuit model of an SRR. The total capacitance across the slot between the rings is C, which is divided into two series components corresponding to the two identical halves of the resonator ($C = 2\pi r_0 C'$, where C' is the per-unit-length capacitance along the slot and $r_0 = r_{ext} - c - d/2$ is the mean radius). The SRR behaves as an externally driven LC resonator

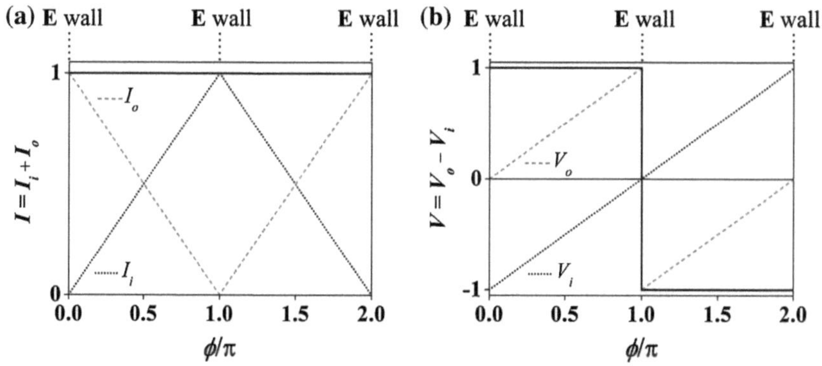

Fig. 2.19 Angular dependence of the quasistatic current and voltage in the SRR; the inner ring is cut at $\phi = 0$. The subscripts i and o stand for the inner and outer rings, respectively. **a** Normalized current distribution; the total current in both rings taken as a summation is constant. **b** Normalized voltage distribution; the voltage across the two rings in each half is constant

As long as the perimeter of the SRR is electrically small with regard to a half-wavelength,[39] a quasistatic analysis is plausible which eases the analysis of the resonator. Under the quasistatic approximation, the behavior of the SRR can be modeled as a self-resonant closed LC circuit as that shown in Fig. 2.18 that may couple to external electromagnetic fields [38]. The inductor L represents the self-inductance of the SRR and the capacitor C stands for the capacitance between the rings. Analytical expressions for L and C have been reported in the literature [3, 38]. In this quasistatic model, the capacitance associated to the slits of the rings is neglected, since most of the electrostatic energy is located between the rings [13]. It is worth mentioning that a more detailed circuit model with distributed elements and taking into account the gap capacitance was later reported [41, 42]. The two gap capacitances, produced by the accumulation of electric charges, can be regarded as being added in parallel to the inter-ring capacitance. It was shown that such a distributed model converges to the quasistatic model when the gap capacitances are neglected and the electrical length of the SRR is small.

Figure 2.19 shows the quasistatic distribution of voltage and current. On each ring, currents vanish at the slits because the gap capacitance is disregarded, whereas

[39]The overall size of the SRR is typically about one-tenth of the wavelength [3].

currents attain maximum values at the center of the rings (namely, $\phi = 0$ for the outer ring and $\phi = \pi$ for the inner ring). Given that the SRR is electrically small, currents on each ring are supposed to vary linearly midway between these extreme values (this results in a triangular current distribution). Nonetheless, the total current on the SRR taken as the sum of the individual currents is uniform around its perimeter. Hence, the whole SRR behaves as a small closed loop of uniform current whose self-inductance can be calculated as the average inductance of the two rings, i.e. as the inductance of a single closed loop with identical width to that of the individual rings (c) and identical mean radius (r_0) [38]. With respect to the voltage distribution, it also exhibits a linear variation in each ring. However, the voltage difference across the slots is constant in each half of the SRR, and the slot between the rings may be modeled by two series capacitances (associated to each half) that store the same amount of charge (but of opposite sign) at both sides of the slot. As expected, the voltage distribution is anti-symmetric with regard to the symmetry axis ($\phi = 0$ and π). Therefore, the voltage is zero at the symmetry plane since a virtual short circuit (an electric wall) appears along this plane.

It is also interesting to analyze the SRR fundamental resonance from the theory of coupled resonators [18]. When the mean circumference of a closed ring resonator is equal to an integral multiple of the wavelength, resonance is established [33]. By cutting the ring by a split, the resulting resonator is essentially a distributed open-circuited half-wavelength resonator with a (weak) loading capacitance at the open end. Since the inner and outer rings of the SRR are not identical, their self-resonances are different from each other. By coupling these half-wavelength rings, the lower resonance frequency (associated to the outer ring) shifts downward, giving rise to quasistatic conditions as a first-order approximation (see Fig. 2.20). For such a reason, the first (or lowest) resonance frequency is referred to as the *fundamental* or *quasistatic* resonance frequency. Obviously, SRRs (and all the other considered resonators) exhibit higher-order resonances that can be excited at higher frequencies (Fig. 2.20). These higher-order resonant modes (beyond the scope of this thesis) are

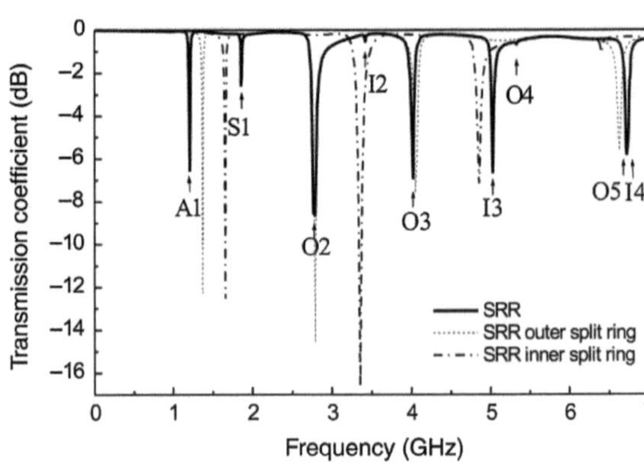

Fig. 2.20 Quasistatic (identified as A1) and some dynamic resonance frequencies of an SRR. The resonances correspond to transmission zeros in the transmission coefficient of an SRR-loaded microstrip line. Reprinted with permission from [43]

referred to as *dynamic* resonances, since non-uniform total currents are induced and the particle is thus no longer electrically small [3, 43].

2.3.2 Double-Slit Split-Ring Resonator (DS-SRR)

The double-slit split-ring resonator (DS-SRR) is derived from the original topology of the SRR by introducing an additional cut in each ring and by rotating one of the rings 90°, as shown in Fig. 2.21a [44]. This particle is bisymmetric, which means that it possesses two orthogonal symmetry planes, resulting invariant by inversion. As Fig. 2.21b depicts, the two symmetry planes behave as electric walls at resonance and, in virtue of their orthogonality, the appearance of a net electric dipole is prevented. As a result, the fundamental resonance of the DS-SRR cannot be electrically excited with uniform fields. Contrarily, an axial (z-direction) time-harmonic magnetic field can establish resonance inducing a magnetic dipole moment. Therefore, this particle does not exhibit cross polarization.

In comparison to the original SRR, by applying a quasistatic analysis, the DS-SRR has almost the same inductance but four times smaller capacitance (provided that the overall dimensions of the resonators are the same). The reduction in the total capacitance arises from the series connection of the four capacitances, each of them corresponding to each quadrant (see Fig. 2.22). As a consequence, the resonance frequency of the DS-SRR is about twice that of the SRR, and the electrical size as well (in fact this is likely to be expected because the length of the individual split

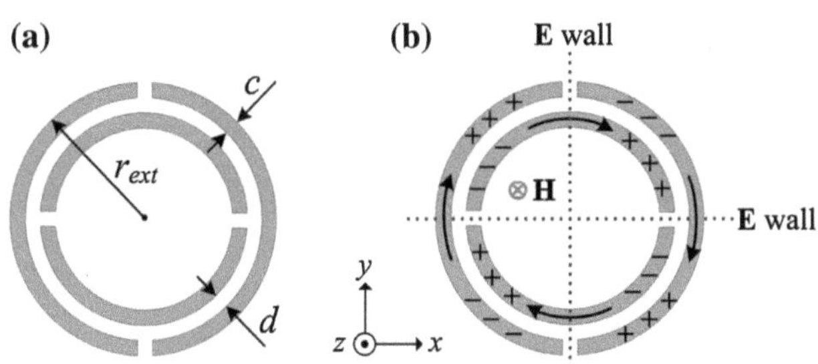

Fig. 2.21 Bisymmetric double-slit split-ring resonator (DS-SRR). **a** Circular topology. **b** Fundamental resonance: orientation of the polarization fields, sketch of the charge distribution and currents, and boundary conditions at the symmetry planes

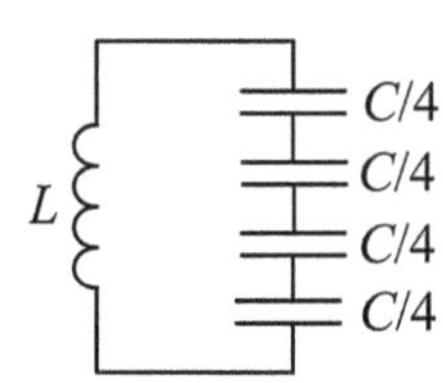

Fig. 2.22 Equivalent circuit model of a DS-SRR. The total capacitance, $C = 2\pi r_0 C'$, is divided into the four quadrants

rings has been decreased by a factor of two). Accordingly, although the equivalent circuit model of the DS-SRR is formally the same as that of the SRR, the description of the electrical behavior of the particle by the model is less accurate for the DS-SRR.

2.3.3 Folded Stepped-Impedance Resonator (FSIR)

Stepped-impedance resonators (SIRs) are common planar building blocks in microwave engineering [45]. SIRs are essentially open-ended resonators where the width of the strip is varied abruptly. It is well known that a tri-section SIR can be made electrically small (as compared to a uniform half-wavelength resonator) by narrowing the central section and widening the external ones. Further miniaturization can be achieved by folding the SIR, as depicted in Fig. 2.23a. The fundamental resonance frequency of the folded SIR (FSIR) can be excited by time-harmonic electric (y-direction) and/or magnetic (z-orientation) fields. Hence, cross-polarization effects are present, and both an electric dipole and a magnetic dipole are induced. At the plane of symmetry, an electric wall boundary condition is established. Figure 2.23b illustrates all the above-cited properties at resonance.

In terms of electric circuit parameters, basically, the inductive component of the particle is directly related to the inductance of the narrow strip (w_2), while its capacitance is provided by the wide sections (w_1) and the gap (s). Thus, an FSIR may be regarded as a planar implementation of a capacitively-loaded loop. As a consequence, the equivalent circuit model of an FSIR (see Fig. 2.24) is qualitatively the same as that of an SRR.

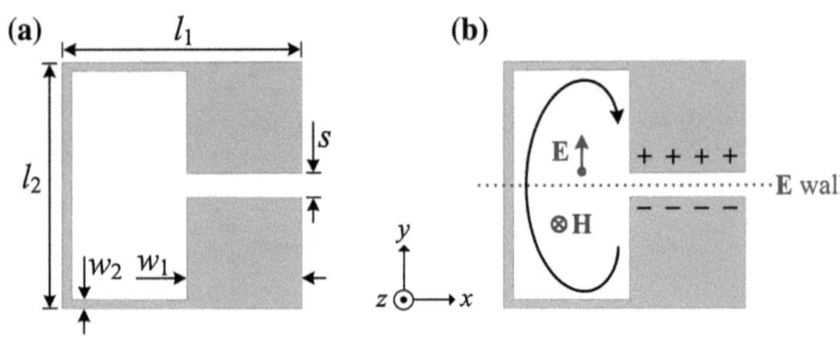

Fig. 2.23 Symmetric folded stepped-impedance resonator (FSIR). **a** Square topology. **b** Fundamental resonance: orientation of the polarization fields, sketch of the charge distribution and currents, and boundary conditions at the symmetry plane

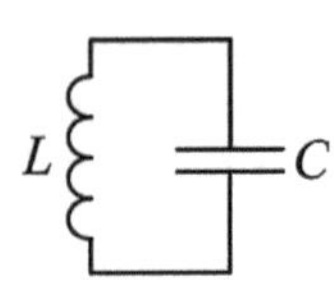

Fig. 2.24 Equivalent circuit model of an FSIR

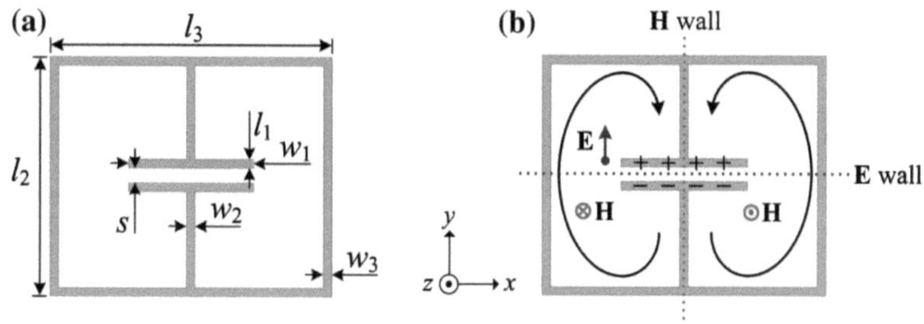

Fig. 2.25 Bisymmetric electric inductive-capacitive (ELC) resonator. **a** Rectangular-shaped topology. **b** Fundamental resonance: orientation of the polarization fields, sketch of the charge distribution and currents, and boundary conditions at the symmetry planes. Since in this thesis the ELC is magnetically coupled to transmission lines, the resonator excitation by contra-directional magnetic fields is also illustrated

2.3.4 Electric Inductive-Capacitive (ELC) Resonator

The so-called electric-field-coupled inductive-capacitive (ELC) resonator (or simply electric-*LC* resonator) was proposed by Schurig et al. [46] as an alternative to metallic wires for the implementation of negative permittivity (ENG) media. As illustrated in Fig. 2.25a, the ELC resonator is a bisymmetric planar structure consisting of a pair of mirrored capacitively-loaded metallic loops in contact. Clearly, this particle can be viewed as two combined (merged) FSIRs.

As indicated in Fig. 2.25b, besides an electric wall, the ELC resonator exhibits a *magnetic wall* in an orthogonal plane at the fundamental resonance. At this resonance, the current flows through the central branch so that the (instantaneous) current is clockwise in one loop and counterclockwise in the other one, giving rise to a net displacement current across the gap. In consequence, the opposite magnetic moments originated by the two current loops cancel out and prevent the presence of a net axial magnetic dipole moment (z-direction). For such a reason this resonator cannot be driven by means of a uniform axial time-varying magnetic field.[40] On the other hand, an electric dipole moment is still present in the plane of the particle directed along the gap (y-oriented), which means that this resonator can be excited through a uniform time-harmonic electric field applied to that direction. As a result, cross polarization does not affect the ELC resonator, a property that can also be inferred because of its inversion symmetry inherent to its bisymmetry.

The equivalent circuit model of an electrically small ELC resonator is schematically depicted in Fig. 2.26 [46]. As in the FSIR, the narrow section of each loop (w_2 and w_3) is associated to an inductance, while the wide section (w_1) and the gap (s) are modeled by a capacitance. Therefore, the ELC resonator is in essence an

[40]Nevertheless, the particle may be magnetically excited if the applied magnetic fields in the individual loops are in opposite directions to each other. In fact, in this thesis the ELC resonator is magnetically coupled to transmission lines.

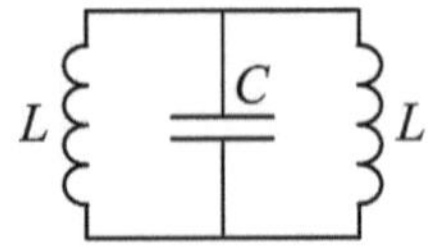

Fig. 2.26 Equivalent circuit model of an ELC resonator. Mesh currents must flow into the same direction through the capacitor for resonance establishment

electrically-coupled *LC* resonator (i.e. an electric resonator) under uniform fields, as its terminology suggests.

2.3.5 Complementary Resonators

The complementary counterparts of the resonators analyzed in the preceding subsections are presented here. As is well known, two complementary structures are defined as those where one is obtained from the other by exchanging the apertures (slots) for the solid (metallic) parts of a plane [47]. Therefore, when complementary structures are combined, they form a single infinite solid screen with no overlaps. As suggested in the literature [48], to avoid ambiguity, a resonator made of electrically conducting strips is referred here to as a *strip resonator*, whereas a resonator obtained by cutting slots in a metallic surface as a *slot resonator*.

With a view to analyzing the behavior of resonators in the form of slots, the concepts of duality and complementarity are invoked as often [47, 48]. Electromagnetic duality (often referred to as Babinet's Principle) means that Maxwell's equations possess duality in electric and magnetic quantities; the mathematical solution for an electromagnetic field is identical to that for its dual one obtained by a suitable interchange of electric and magnetic quantities. The dual of a strip resonator can be mathematically obtained by the definition of fictitious magnetic conductors. In practice, a physical dual is implemented by a slot resonator. Thus, strip and slot resonators are the physical implementation of dual problems. It follows that, the electromagnetic fields of complementary structures are identical with the exception of a constant and that the electric and magnetic fields are interchanged. Therefore, the electric/magnetic field distribution of a planar slot structure is the same as the magnetic/electric field distribution of a planar strip structure (the sign of the dual field, however, must change on each side of the slot structure in order to satisfy the boundary conditions in the plane of the particle). It is also well known that the impedance and admittance of complementary structures are proportional, and accordingly their resonance frequencies are identical. Another interesting property derived from duality is that complementary periodic structures have exactly the same dispersion curve [9]. A case study on this matter will be addressed in Sect. 2.4.

It should be pointed out that duality in complementary structures strictly applies to infinite, planar, zero thickness, and perfectly conducting electric screens [47].

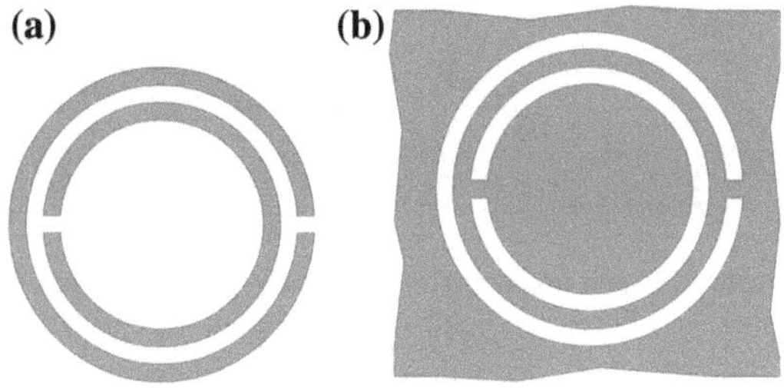

Fig. 2.27 Complementary resonators. **a** SRR and **b** CSRR

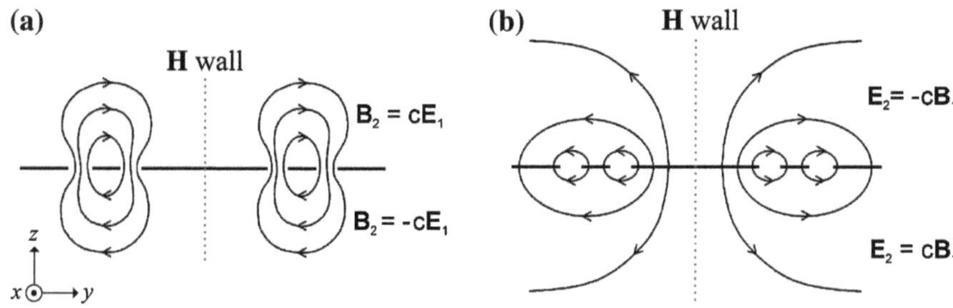

Fig. 2.28 Sketch of the magnetic **a** and electric **b** field distributions at the CSRR fundamental resonance frequency. The field distributions of the CSRR and SRR are dual, *c* being the velocity of light. Reprinted with permission from [21]

Although in practice such ideal conditions cannot be fulfilled, for very thin good conductors and metallic planes substantially larger than the wavelength and the slot region, Babinet's Principle predicts the actual behavior to a high degree [37]. One of the most relevant discrepancies due to deviations from ideal conditions is a shift in the resonance frequency.

Let us consider as an illustration the complementary screen of the SRR, i.e. the complementary split-ring resonator (CSRR) [22], as shown in Fig. 2.27.[41] The electromagnetic field distribution at the fundamental resonance of the CSRR is sketched in Fig. 2.28. It is apparent that the electric and magnetic fields in the vicinity of the CSRR are the dual fields of those of the SRR (see Fig. 2.17), since the roles of the electric and magnetic fields are interchanged. Thus, at the symmetry plane of the CSRR, a magnetic wall arises. The induced electric and magnetic dipoles are also dual, as illustrated in Fig. 2.29 [23]. While the SRR behaves as an axial magnetic dipole and an electric dipole in the plane of the rings, the CSRR (when is seen from one side) exhibits an axial electric dipole and a magnetic dipole in its plane. Note that the dipoles point toward opposite directions on the two sides ($z < 0$ and $z > 0$) of the CSRR. Therefore, the net electric and magnetic dipoles generated by the CSRR become zero. This ensures that both the total magnetic polarization parallel to the screen (y-axis) and the total electric polarization perpendicular to the screen (z-axis)

[41] Despite the nomenclature, the SRR is the complementary of the CSRR as well.

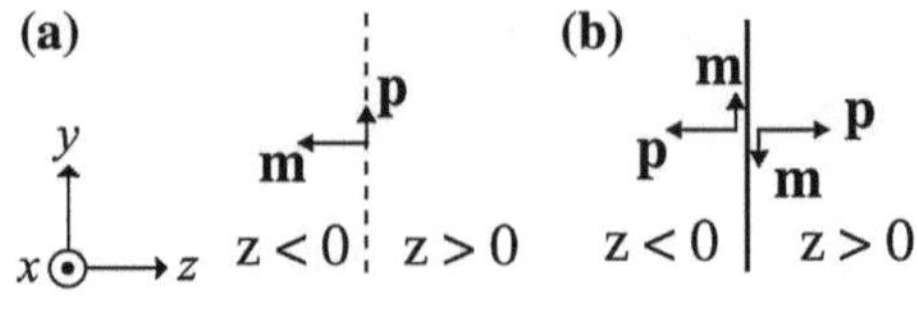

Fig. 2.29 Induced electric (p) and magnetic (m) dipoles by the **a** SRR and the **b** CSRR [23]. The resonators lie in the plane $z = 0$. No net dipoles are induced in the CSRR seen from both sides

vanish, as it must be for a metallic plane (electric currents in a plane can produce only net tangential electric moments and normal magnetic moments) [3]. As a consequence, when referring to slot resonators, uniform and temporally varying fields are applied to only one side. The CSRR can thus be excited by an axial electric field and/or a magnetic field polarized in the direction orthogonal to its symmetry plane, as depicted in Fig. 2.30a. Hence, cross-polarization effects are also present in the CSRR, the electric field being the main driving mechanism [21].

Figure 2.30 shows the results after applying duality to all the considered resonators. The complementary of the DS-SRR is called double-slit complementary split-ring resonator (DS-CSRR) [44]. Like the SIR, its complementary counterpart [49] has been extensively used in the last decade in the microwave community, and it is referred to as dumbbell-shaped defected ground structure (DGS)[42] here, as it is often called in the literature. Neither the unfolded nor the folded dumbbell DGS is used in the present thesis, but it is included here for completeness. The physical dual of the ELC resonator has been called magnetic-LC (MLC) resonator for coherence throughout this thesis, meaning that uniform fields are able to excite its fundamental resonance only by magnetic fields (the MLC is therefore a magnetic resonator). An alternative denomination in the literature is complementary electric-LC (CELC) resonator [50].

Finally, it may be shown that electromagnetic duality implies circuit theory duality. Consequently, any planar structure and its complementary screen are dual in circuit theory as well [51]. Thus, the equivalent circuit models of slot resonators may be inferred by the dual circuits of the models corresponding to the strip resonators. By applying circuit duality, inductances are replaced with capacitances, and series-connected circuit elements become parallel-connected circuit elements. The equivalent circuit models for the CSRR and the DS-CSRR are represented in Fig. 2.31 [21]. The capacitance of the CSRR may be calculated as the capacitance of a disk of radius $r_0 - c/2$ ($r_0 = r_{ext} - c - d/2$) surrounded by a metal plane at a distance c of its edge (analytical expressions have been derived [21]). The CSRR inductance is given by the parallel combination of the two inductances connecting the inner disk to the outer metallic region of the CSRR. Such inductances may be calculated as the inductance of a CPW with a strip width d and slot width c. A circuit model of the MLC resonator was developed by Naqui et al. in [52], while the circuit of the folded dumbbell-shaped DGS can be readily derived from that circuit.

[42] A *slot* resonator is sometimes referred to as a *defected ground structure* (DGS) or *patterned ground structure* (PGS) when it is etched in the ground plane.

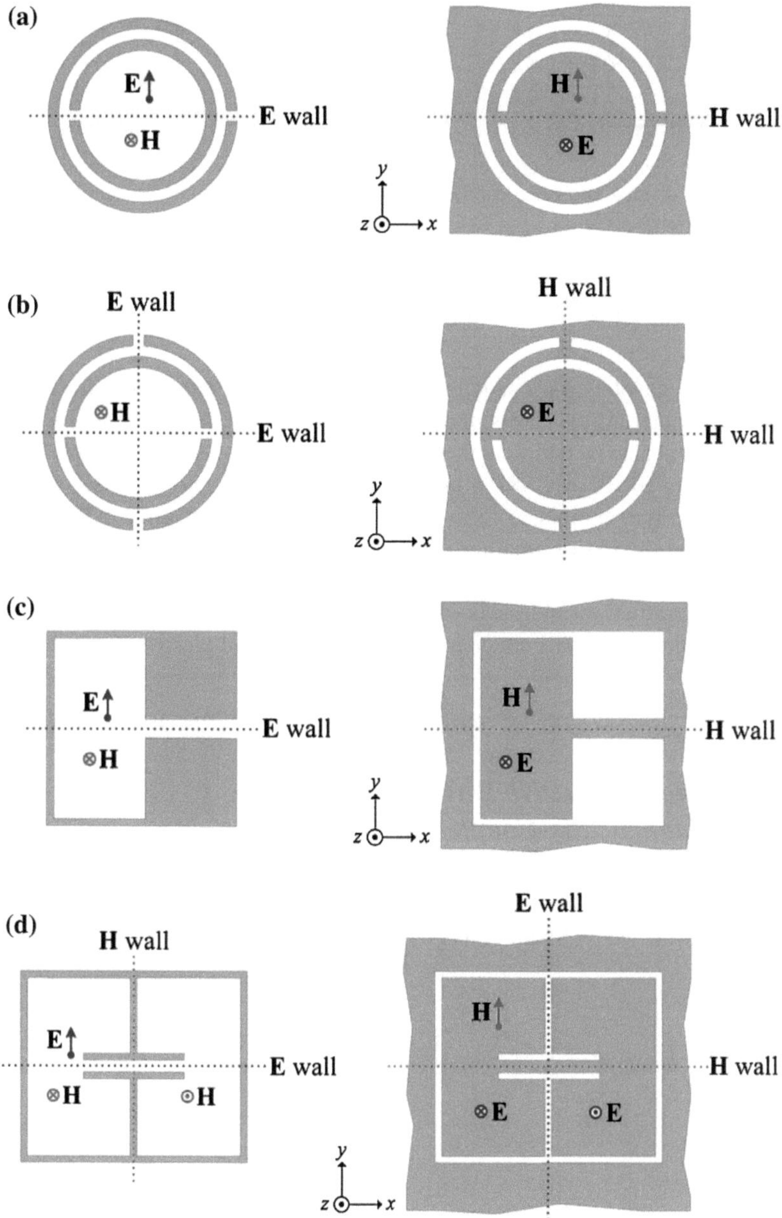

Fig. 2.30 Polarization fields and boundary conditions at the symmetry planes at the fundamental resonance in complementary split-ring resonators by applying duality

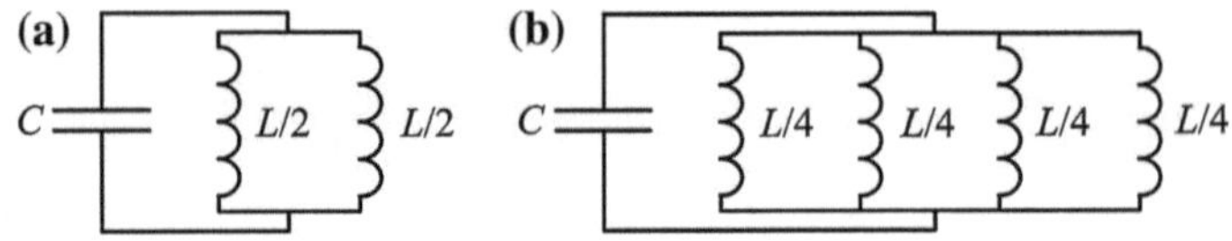

Fig. 2.31 Equivalent circuit model of a **a** CSRR and a **b** DS-CSRR. The total inductance, $L = 2\pi r_0 L'$, is divided into equal parallel contributions, where L' is the per-unit-length inductance of an equivalent CPW

2.4 Magneto- and Electro-Inductive Waves

This section is devoted to propagating structures based on inter-resonator coupling. Indeed, metamaterial resonators may be excited not only, for instance, by a transmission line, but also by coupling them. Without loss of generality, imagine periodic chains of either SRRs or CSRRs, as depicted in Fig. 2.32. Since these two arrays complement to one another to form a solid screen, both structures exhibit numerically the same dispersion relation [9], as mentioned in Sect. 2.3.5. However, it is apparent from duality that the nature of the propagation in the two arrays must somehow interchange dual variables. As will be argued, planar arrays of SRRs and CSRRs are able to guide backward *magneto-inductive waves* and *electro-inductive waves*, respectively.

2.4.1 *Magneto-Inductive Waves in Arrays of Magnetically-Coupled Resonators*

Magneto-inductive waves (MIWs) were first reported and studied in the frame of metamaterial science by Shamonina et al. [53, 54], and many works have been pub-

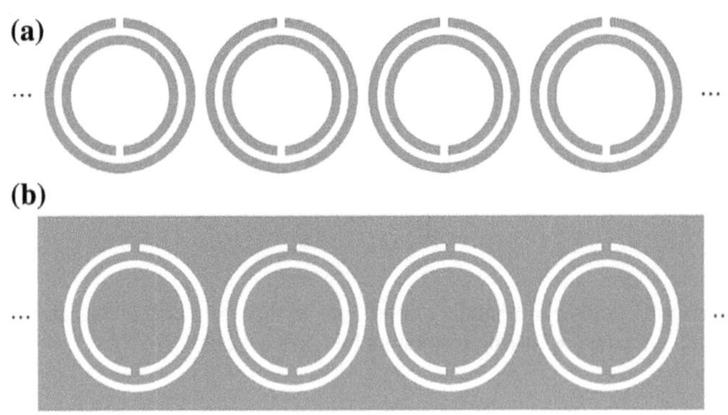

Fig. 2.32 Periodic array of coupled **a** SRRs and **b** CSRRs through which dual magneto- and electro-inductive waves may arise, respectively

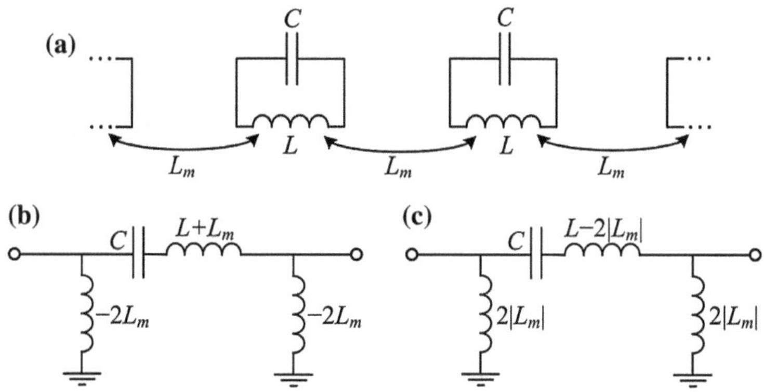

Fig. 2.33 Equivalent circuit of a periodic magneto-inductive waveguide assuming magnetic interaction only between adjacent resonators. **a** Circuit with mutual inductance. Unit-cell transformed circuits (see Appendix B.2) for an **b** axial and a **c** coplanar arrangement of resonators supporting forward and backward waves, respectively

lished subsequently [5, 55, 56]. MIWs are a form of guided waves which propagate through an array of magnetically-coupled resonators that acts as a waveguide. A lumped-element circuit representation of a periodic magneto-inductive waveguide is illustrated in Fig. 2.33a. Each resonator is modeled by an inductance L and a capacitance C, and is coupled to its nearest neighbor[43] by a mutual inductance L_m. Thus, the term *magneto-inductive wave* originates from the fact that such waves arise due to voltages induced by magnetic coupling between the resonant elements [57]. The propagation of MIWs occurs within a narrow passband around the resonance frequency of the isolated elements, i.e. $\omega_0 = 1/\sqrt{LC}$, and the associated bandwidth is dependent on how strongly the resonators are coupled to each other. The stronger the interaction between the elements, the wider the passband of the MIWs [57]. Both forward and backward MIWs may exist, depending on the sign of the mutual inductance between resonators (see Fig. 2.33b, c). While mutual inductance is positive in an axial configuration, it is negative when the resonators are arranged in a coplanar configuration since the magnetic field originating in one resonator crosses another one in the opposite direction [54].

MIWs are expected to be present in whatever chain of resonant elements that can be described by LC resonant tanks, as long as the coupling between them is magnetic [54, 58] or dominantly magnetic in the presence of negligible mixed magnetoelectric coupling [3]. The first magneto-inductive waveguide was studied in the form of a theoretical periodic array of magnetically-coupled capacitively-loaded loops, as can be seen in Fig. 2.34a for a coplanar arrangement [54]. The corresponding dispersion diagram assuming that coupling is restricted to the nearest neighbors is shown in

[43] The first-neighbor approximation assumes that the field strength declines so rapidly away from a resonator that it is too small beyond the nearest neighbor. The accuracy of this first-order coupling is dependent on the resonant elements [3, 5] and the spacing between resonators.

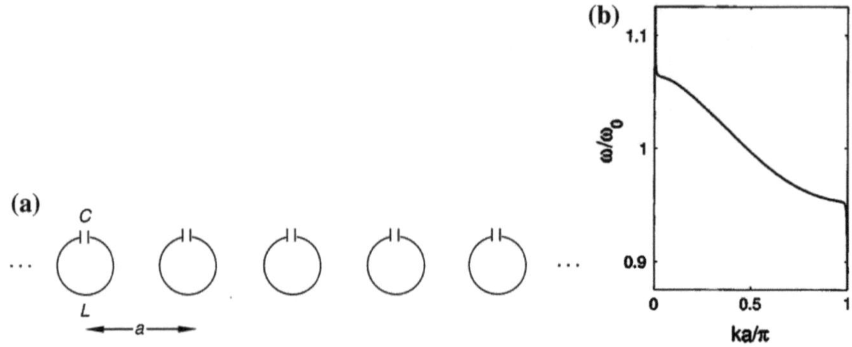

Fig. 2.34 Backward magneto-inductive waves in an array of coplanar capacitively-loaded loops. **a** Theoretical layout reprinted with permission from [59]. **b** Dispersion relation with nearest neighbor coupling reprinted with permission from [54]

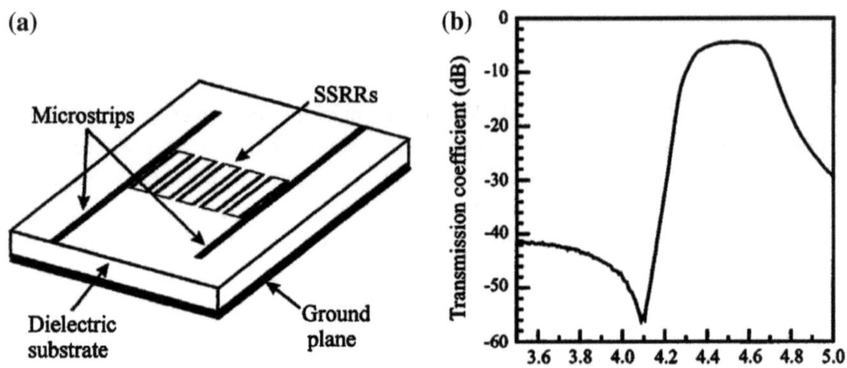

Fig. 2.35 Backward magneto-inductive waves in an array of split-ring resonators in coplanar configuration in microstrip technology. **a** Layout; the input/output SSRRs are coupled to input/output microstrip lines. **b** Measured transmission coefficient. Reprinted with permission from [58]

Fig. 2.34b, where the backward wave nature is clearly visible. Shortly after, MIWs were also excited in a periodic arrangement of split squared ring resonators (SSRRs), as shown in Fig. 2.35 [58]. This type of split-ring resonator (not described in Sect. 2.3) is not symmetric, but it does not suffer from cross-polarization effects as it exhibits inversion symmetry.

It is also worth mentioning that MIWs belong to the category of slow waves, since its phase velocity may be orders of magnitude smaller than the velocity of an electromagnetic wave propagating in the same medium [3, 6].[44] Since slow-wave structures are of interest in device miniaturization, and bandpass characteristics are useful in signal filtering, MIWs have found many diverse applications. Some

[44]Slow-wave effects are usually accounted for by the resulting phase velocity reduction factor (or slow-wave factor).

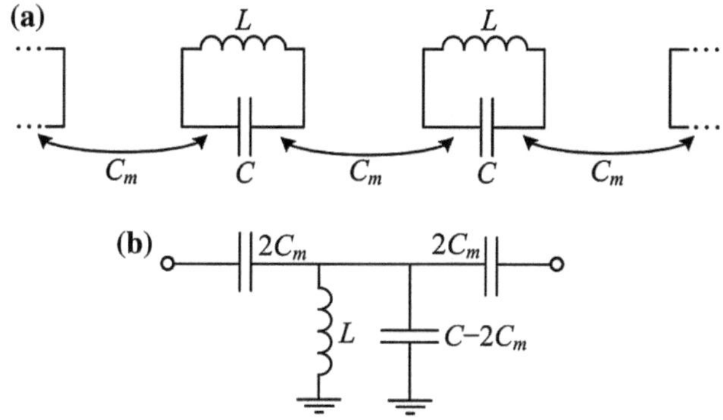

Fig. 2.36 Equivalent circuit of a periodic electro-inductive waveguide with electric interaction only between adjacent resonators. **a** Circuit with mutual capacitance. **b** Unit-cell transformed circuit model (see Appendix B.1) for a coplanar arrangement of resonators where wave propagation is backward

examples are power dividers [54], delay lines [58], passive chipless tags for radio-frequency identification (RFID) [60], and phase shifters [61].

2.4.2 Electro-Inductive Waves in Arrays of Electrically-Coupled Resonators

Electro-inductive waves (EIWs) can be regarded as the dual counterpart of magneto-inductive waves [62–65].[45] Therefore, EIWs propagate through a waveguide whose guiding mechanism is based on electric interaction between neighboring resonant elements, in parallelism (or duality) to MIWs. The lumped-element equivalent circuit model of a periodic electro-inductive waveguide with the first-neighbor approximation is depicted in Fig. 2.36a, where the mutual capacitance C_m accounts for the inter-resonator electric coupling. The resulting unit-cell equivalent-circuit model for a planar arrangement of electrically-coupled resonators, which supports backward propagation, is shown in Fig. 2.36b. As can be seen, for planar structures, the circuit of electro-inductive waveguides is dual to the circuit modeling magneto-inductive waveguides (Figs. 2.33c and 2.36b). This is not surprising because electromagnetic duality involves circuit duality, as was asserted in Sect. 2.3.5. Therefore, the dispersion relations inferred from the equivalent circuit models of magneto- and electro-inductive waveguides are equal to each other if the circuits are numerically dual. The meaningful difference lies in the nature (MIW- or EIW-related) of the propagating band.

[45]However, propagation through electrically-coupled metallic rods have been known for a long time [66].

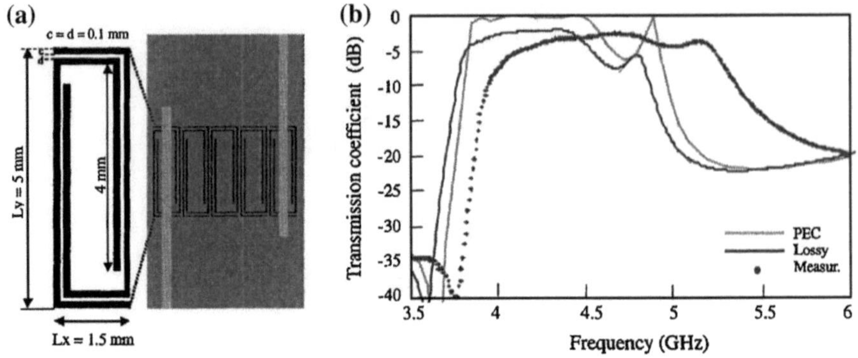

Fig. 2.37 Electro-inductive waves in an array of complementary split-ring resonators in coplanar configuration in microstrip implementation. **a** Layout and **b** simulated and measured transmission coefficient. Reprinted with permission from [62]

The first reported electro-inductive waveguide was implemented in microstrip technology using the complementary split squared ring resonator (CSSRR), that is, the complementary version of the SSRR [62]. Figure 2.37 shows the layout of the structure as well as the transmission coefficient. As expected, this waveguide behaves in a similar way as its dual counterpart[46] [58] (Fig. 2.35).

References

1. C. Caloz, T. Itoh, *Electromagnetic Metamaterials: Transmission Line Theory and Microwave Applications* (Wiley, New York, 2005)
2. N. Engheta, R.W. Ziolkowski, *Metamaterials: Physics and Engineering Explorations* (Wiley, New York, 2006)
3. R. Marqués, F. Martín, M. Sorolla, *Metamaterials with Negative Parameters: Theory, Design and Microwave Applications* (Wiley, New York, 2008)
4. G.V. Eleftheriades, K.G. Balmain, *Negative-Refraction Metamaterials: Fundamental Principles and Applications* (Wiley, New York, 2005)
5. L. Solymar, E. Shamonina, *Waves in Metamaterials* (Oxford University Press, New York, 2009)
6. F. Martín, *Artificial Transmission Lines for RF and Microwave Applications* (Wiley, New York, 2015)
7. S. Ramo, J.R. Whinnery, T. Van Duzer, *Fields and Waves in Communication Electronics* (Wiley, New York, 1965)
8. C.A. Balanis, *Advanced Engineering Electromagnetics* (Wiley, New York, 1989)
9. R.E. Collin, *Foundations for Microwave Engineering*, 2nd edn. (McGraw-Hill, New York, 1992)
10. D.M. Pozar, *Microwave Engineering*, 3rd edn. (Wiley, New York, 2005)
11. V.G. Veselago, The electrodynamics of substances with simultaneously negative values of ϵ and μ. Physics-Uspekhi **10**(4), 509–514 (1968)

[46]In a strict sense, however, multilayer structures cannot be dual.

12. D.R. Smith, W.J. Padilla, D.C. Vier, S.C. Nemat-Nasser, S. Schultz, Composite medium with simultaneously negative permeability and permittivity. Phys. Rev. Lett. **84**(18), 4184–4187 (2000)
13. J.B. Pendry, A.J. Holden, D.J. Robbins, W.J. Stewart, Magnetism from conductors and enhanced nonlinear phenomena. IEEE Trans. Microw. Theory Tech. **47**(11), 2075–2084 (1999)
14. A.K. Iyer, G.V. Eleftheriades, Negative refractive index metamaterials supporting 2-D waves, in *IEEE MTT-S International Microwave Symposium Digest*, vol. 2 (Seattle, WA, USA, 2002). pp. 1067–1070
15. A.A. Oliner, A periodic-structure negative-refractive-index medium without resonant elements, in *IEEE-AP-S USNC/URSI National Radio Science Meeting* (San Antonio, TX, USA, 2002). p. 41
16. C. Caloz, T. Itoh, Application of the transmission line theory of left-handed (LH) materials to the realization of a microstrip "LH line", in *IEEE-AP-S USNC/URSI National Radio Science Meeting*. vol. 2 (2002). pp. 412–415
17. C. Caloz, T. Itoh, Novel microwave devices and structures based on the transmission line approach of meta-materials, in *IEEE MTT-S International Microwave Symposium Digiet*, vol. 1 (Philadelphia, PA, USA, 2003). pp. 195–198
18. J.S. Hong, M.J. Lancaster, *Microstrip Filters for RF/Microwave Applications* (Wiley, New York, 2001)
19. A. Grbic, G.V. Eleftheriades, Experimental verification of backward-wave radiation from a negative refractive index metamaterial. J. Appl. Phys. **92**(10), 5930–5935 (2002)
20. F. Martín, J. Bonache, F. Falcone, M. Sorolla, R. Marqués, Split ring resonator-based left-handed coplanar waveguide. Appl. Phys. Lett. **83**(22), 4652–4654 (2003)
21. J.D. Baena, J. Bonache, F. Martín, R.M. Sillero, F. Falcone, T. Lopetegi, M.A.G. Laso, J. García-García, I. Gil, M.F. Portillo, M. Sorolla, Equivalent-circuit models for split-ring resonators and complementary split-ring resonators coupled to planar transmission lines. IEEE Trans. Microw. Theory Tech. **53**(4), 1451–1461 (2005)
22. F. Falcone, T. Lopetegi, J.D. Baena, R. Marqués, F. Martín, M. Sorolla, Effective negative-ϵ; stopband microstrip lines based on complementary split ring resonators. IEEE Microw. Wirel. Compon. Lett. **14**(6), 280–282 (2004)
23. F. Falcone, T. Lopetegi, M. Laso, J. Baena, J. Bonache, M. Beruete, R. Marqués, F. Martín, M. Sorolla, Babinet principle applied to the design of metasurfaces and metamaterials. Phys. Rev. Lett. **93**(197401) (2004)
24. F. Aznar, J. Bonache, F. Martín, Improved circuit model for left-handed lines loaded with split ring resonators. Appl. Phys. Lett. **92**(043512) (2008)
25. J. Bonache, M. Gil, O. García-Abad, F. Martín, Parametric analysis of microstrip lines loaded with complementary split ring resonators. Microw. Opt. Technol. Lett. **50**(8), 2093–2096 (2008)
26. F. Martín, F. Falcone, J. Bonache, R. Marqués, M. Sorolla, Miniaturized coplanar waveguide stop band filters based on multiple tuned split ring resonators. IEEE Microw. Wirel. Compon. Lett. **13**(12), 511–513 (2003)
27. I.J. Bahl, P. Bhartia, *Microwave Solid State Circuit Design* (Wiley, New York, 2003)
28. F. Aznar, M. Gil, J. Bonache, L. Jelinek, J.D. Baena, R. Marqués, F. Martín, Characterization of miniaturized metamaterial resonators coupled to planar transmission lines through parameter extraction. J. Appl. Phys. **104**(114501) (2008)
29. J. Bonache, M. Gil, I. Gil, J. García-García, F. Martín, On the electrical characteristics of complementary metamaterial resonators. IEEE Microw. Wirel. Compon. Lett. **16**(10), 543–545 (2006)
30. L. Brillouin, *Wave Propagation in Periodic Structures: Electric Filters and Crystal Lattices* (McGraw-Hill, New York, 1946)
31. F. Aznar, M. Gil, J. Bonache, F. Martín, Modelling metamaterial transmission lines: a review and recent developments. Opto Electron. Rev. **16**(3), 226–236 (2008)
32. F. Bilotti, L. Vegni, From metamaterial-based to metamaterial-inspired miniaturized antennas: A possible procedure and some examples, in *Proceedings of URSI XXIX General Assembly* (2008) pp. 1–4

33. K. Chang, L.-H. Hsieh, *Microwave Ring Circuits and Related Structures* (Wiley, New York, 2004)
34. J. Martel, R. Marqués, F. Falcone, J. Baena, F. Medina, F. Martín, M. Sorolla, A new LC series element for compact bandpass filter design. IEEE Microw. Wirel. Compon. Lett. **14**(5), 210–212 (2004)
35. A. Vélez, F. Aznar, J. Bonache, M. Velázquez-Ahumada, J. Martel, F. Martín, Open complementary split ring resonators (OCSRRs) and their application to wideband CPW band pass filters. IEEE Microw. Wirel. Compon. Lett. **19**(4), 197–199 (2009)
36. W.N. Hardy, L.A. Whitehead, Split-ring resonator for use in magnetic resonance from 200–2000 MHz. Rev. Sci. Instrum. **52**(2), 213–216 (1981)
37. C.A. Balanis, *Antenna Theory: Analysis and Design*, 2nd edn. (Wiley, New York, 1997)
38. R. Marqués, F. Medina, R. Rafii-El-Idrissi, Role of bianisotropy in negative permeability and left-handed metamaterials. Phys. Rev. B **65**(144440) (2002)
39. P. Gay-Balmaz, O.J.F. Martin, Electromagnetic resonances in individual and coupled split-ring resonators. J. Appl. Phys. **92**(5), 2929–2936 (2002)
40. N. Katsarakis, T. Koschny, M. Kafesaki, E. Economou, C. Soukoulis, Electric coupling to the magnetic resonance of split ring resonators. Appl. Phys. Lett. **84**(15), 2943–2945 (2004)
41. M. Shamonin, E. Shamonina, V. Kalinin, L. Solymar, Properties of a metamaterial element: analytical solutions and numerical simulations for a singly split double ring. J. Appl. Phys. **95**(7), 3778–3784 (2004)
42. M. Shamonin, E. Shamonina, V. Kalinin, L. Solymar, Resonant frequencies of a split-ring resonator: analytical solutions and numerical simulations. Microw. Opt. Technol. Lett. **44**(2), 133–136 (2005)
43. J. García-García, F. Martín, J.D. Baena, R. Marqués, L. Jelinek On the resonances and polarizabilities of split ring resonators. J. Appl. Phys. **98**(033103) (2005)
44. R. Marqués, J. Baena, J. Martel, F. Medina, F. Falcone, M. Sorolla, F. Martín, Novel small resonant electromagnetic particles for metamaterial and filter design, in *International Conference on Electromagnetics in Advanced Applications (ICEAA '03)* (Torino, Italy, 2003). pp. 439–442
45. M. Makimoto, S. Yamashita, Compact bandpass filters using stepped impedance resonators. Proc. IEEE **67**(1), 16–19 (1979)
46. D. Schurig, J. Mock, D. Smith, Electric-field-coupled resonators for negative permittivity metamaterials. Appl. Phys. Lett. **88**(041109) (2006)
47. H. Booker, Slot aerials and their relation to complementary wire aerials (Babinet's principle). J. IEE Part IIIA **93**(4), 620–626 (1946)
48. R. King, G.H. Owyang, Complementarity in the study of transmission lines. IEEE Trans. Microw. Theory Tech. **8**(2), 172–181 (1960)
49. C.-S. Kim, J.-S. Park, D. Ahn, J.-B. Lim, A novel 1-D periodic defected ground structure for planar circuits. IEEE Microw. Guided Wave Lett. **10**(4), 131–133 (2000)
50. T. Hand, J. Gollub, S. Sajuyigbe, D. Smith, S. Cummer, Characterization of complementary electric field coupled resonant surfaces. Appl. Phys. Lett. **93**(212504) (2008)
51. W. Getsinger, Circuit duals on planar transmission media, in *IEEE MTT-S International Microwave Symposium Digest* (Boston, MA, USA, 1983). pp. 154–156
52. J. Naqui, M. Durán-Sindreu, F. Martín, Differential and single-ended microstrip lines loaded with slotted magnetic-LC (MLC) resonators. Int. J. Antennas Propag. **2013**(640514) (2013)
53. E. Shamonina, V.A. Kalinin, K.H. Ringhofer, L. Solymar, Magneto-inductive waveguide. Electron. Lett. **38**(8), 371–373 (2002)
54. E. Shamonina, V.A. Kalinin, K.H. Ringhofer, L. Solymar, Magnetoinductive waves in one, two, and three dimensions. J. Appl. Phys. **92**(10), 6252–6261 (2002)
55. V. Lomanets, O. Zhuromskyy, G. Onishchukov, O. Sydoruk, E. Tatartschuk, E. Shamonina, G. Leuchs, U. Peschel, Interacting waves on chains of split-ring resonators in the presence of retardation. Appl. Phys. Lett. **97**(011108) (2010)
56. I.V. Shadrivov, A.N. Reznik, Y.S. Kivshar, Magnetoinductive waves in arrays of split-ring resonators. Physica B Condens. Matter **394**(2), 180–183 (2007)

57. R.R.A. Syms, E. Shamonina, L. Solymar, Positive and negative refraction of magnetoinductive waves in two dimensions. Europ. Phys. J. B **46**(2), 301–308 (2005)
58. M.J. Freire, R. Marqués, F. Medina, M.A.G. Laso, F. Martín, Planar magnetoinductive wave transducers: theory and applications. Appl. Phys. Lett. **85**(19), 4439–4441 (2004)
59. R.R.A. Syms, E. Shamonina, L. Solymar, Magneto-inductive waveguide devices. IEE Proc. Microw. Antennas Propag. **153**(2), 111–121 (2006)
60. F.J. Herraiz-Martínez, F. Paredes, G. Zamora González, F. Martín, J. Bonache, Printed magnetoinductive-wave (MIW) delay lines for chipless RFID applications. IEEE Trans. Antennas Propag. **60**(11), 5075–5082 (2012)
61. I.S. Nefedov, S.A. Tretyakov, On potential applications of metamaterials for the design of broadband phase shifters. Microw. Opt. Technol. Lett. **45**(2), 98–102 (2005)
62. M. Beruete, F. Falcone, M.J. Freire, R. Marqués, J.D. Baena, Electroinductive waves in chains of complementary metamaterial elements. Appl. Phys. Lett. **88**(083503) (2006)
63. R.R.A. Syms, L. Solymar, A generic approach to boundary reflection in periodic media. Eur. Phys. J. B **54**(2), 169–174 (2006)
64. M. Aznabet, O. El Mrabet, M. Navarro, M. Beruete, F. Falcone, M. Essaaidi, M. Sorolla. Wave propagation properties in stacked SRR/CSRR metasurfaces at microwave frequencies, in *Mediterranean Microwave Symposium (MMS)* (Tangier, Morocco, 2009)
65. M. Beruete, M. Aznabet, M. Navarro-Cía, O. El Mrabet, F. Falcone, N. Aknin, M. Essaaidi, M. Sorolla, Electroinductive waves role in left-handed stacked complementary split rings resonators. Opt. Exp. **17**(3), 1274–1281 (2009)
66. J. Shefer, Periodic cylinder arrays as transmission lines. IEEE Trans. Microw. Theory Tech. **11**(1), 55–61 (1963)

Chapter 3
Advances in Equivalent Circuit Models of Resonator-Loaded Transmission Lines

The present chapter focuses on advances relative to equivalent circuit models of transmission lines loaded with electrically small resonators. An accurate circuit model should include all the elements (a transmission line and an array of resonators in our case) as well as all the inter-element interactions involved. Note that neither mixed magnetoelectric coupling nor inter-resonator coupling was considered in the previous chapter (Sect. 2.2.4). The mixed coupling between the line and the resonators is usually neglected because only the dominant coupling is the one modeled, but it may become important under some resonator orientations. Regarding the coupling between resonators, it may be ignored when the resonators are not closely spaced. By contrast, if the resonators are close to each other, inter-resonator interaction must be taken into account since it influences the frequency response to a considerable extent. In this chapter, mixed coupling in the line-to-resonator interaction and the dominant coupling between contiguous resonators are analyzed, providing unit-cell equivalent circuit models. Additionally, the limits on the synthesis of electrically small stepped-impedance shunt-stubs (SISSs) are also investigated. While the *analysis* is aimed at getting useful electromagnetic information from a particular structure, the inverse process is called *synthesis*. The latter is of paramount importance for design and optimization purposes.

This chapter is divided into three sections. Section 3.1 deals with mixed coupling between the transmission line and the resonators. In Sect. 3.2 resonator-to-resonator coupling effects on the frequency response as well as on the dispersion relation are analyzed. Lastly, Sect. 3.3 reports the proposed method to obtain the physically implementable values of inductance and capacitance using electrically small SISSs.

J. Naqui, *Symmetry Properties in Transmission Lines Loaded with Electrically Small Resonators*, Springer Theses,
DOI 10.1007/978-3-319-24566-9_3

3.1 Line-to-Resonator Magnetoelectric Coupling

As was explained in Sect. 2.3.1, resonators that do not possess inversion symmetry may suffer from cross-polarization effects [1]. This means the resonators may become electrically polarized as a response to an applied magnetic field and vice versa [2, 3], and that they may be magnetically and/or electrically excited, provided the resonators are properly oriented [4, 5]. This section addresses the study of the relative orientation of metamaterial-inspired resonators when they are loading elements in transmission lines, where the resonators are illuminated by non-uniform fields.

Mixed coupling is usually ignored so that the dominant coupling is the only coupling mechanism modeled [6]. However, depending on the resonator orientation in regard to the transmission line, mixed magnetoelectric coupling arises and may become relevant. As will be shown in Chaps. 4–7, the resonator orientation plays a key role in the presented symmetry-based devices. The focus in the present section is on modeling only pairs[1] of split-ring resonators (SRRs) loading coplanar waveguides (CPWs) and single (See footnote 1) complementary split-ring resonators (CSRRs) loading microstrip lines [7]. This suffices to illustrate the modeling of magnetoelectric coupling. Nevertheless, mixed-coupling effects have been also modeled in CPWs loaded with single CSRRs or SRRs (Sect. 4.2.1) [8], and in ELC- and MLC-loaded differential microstrip lines (Sect. 4.2.3) [9].

3.1.1 Coplanar Waveguides Loaded with Pairs of SRRs and CSRR-Loaded Microstrip Lines

As shown in Sect. 2.2.4, equivalent circuit models of SRR-loaded CPWs and CSRR-loaded microstrip lines were proposed by Martín et al. [10, 11] and Baena et al. [6], respectively (the former was revised by Aznar et al. [12]). In these models, mixed coupling is not considered. Since these circuit models are symmetric with regard to the ports, the SRR- and CSRR-loaded lines should also be symmetric when looking into the two ports. This means that there is a symmetry plane midway the ports, and the ports can be interchanged. Certainly, in the literature the slits of the SRRs/CSRRs are usually (but not always) oriented orthogonally to the line axis, as depicted in Fig. 3.1 for $\theta = 0°$. Under these circumstances, these resonator-loaded lines are symmetric with regard to the ports, and the reflection coefficients are indistinguishable ($S_{11} = S_{22}$).[2] Indeed, it has been demonstrated through circuit parameter extraction that these models describe the frequency response of the corresponding structures to a very good approximation [13, 14]. However, as indicated by Naqui et al. [7], if the symmetry of the loaded lines (in regard to the ports) is disrupted by rotating the SRRs/CSRRs, as Fig. 3.1 illustrates for $\theta \neq 90°$, the models as well as the

[1]Quantity of resonators per-unit-cell (placed transversely to the transmission line).

[2]In transmission lines loaded with pairs of symmetric resonators, the longitudinally-oriented mid-plane can also be symmetric (e.g. in the considered SRR-loaded CPWs for $\theta = 90°$).

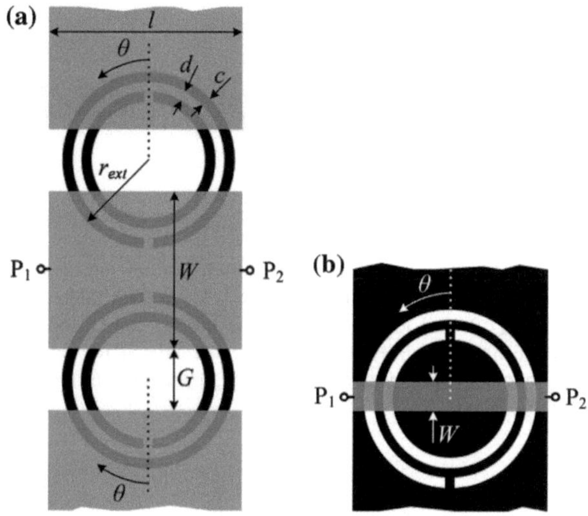

Fig. 3.1 Unit cell topology of **a** a CPW loaded with pairs of SRRs and **b** a microstrip line loaded with CSRRs. The upper metal layer is depicted in *gray*, whereas the lower one in *black*. The rotation angle that defines the slits orientation ranges within $0° \leq \theta \leq 90°$ and, for the particular topology depicted, $\theta = 0°$. Notice that the rotation in **a** is that in which the slot mode cannot be excited regardless of the rotation since symmetry along the longitudinally-oriented midplane is preserved

parameter extraction procedures may fail. For the particular case where the slits are oriented in parallel to the line axis (i.e. $\theta = 90°$, where the structures are symmetric with regard to the midplane extended along the longitudinal direction, but not to midplane extended in the transverse direction), the asymmetry leads to substantially different phases of the two reflection coefficients.[3] As mentioned earlier, symmetry along the longitudinal midplane is of paramount importance in the applications considered in this book.

Hence, the properties and effects due to the rotation of SRRs and CSRRs on the frequency response of SRR-loaded CPWs and CSRR-loaded microstrip lines are investigated in [7]. If the SRR/CSRR slits are aligned orthogonally to the line axis, the SRRs/CSRRs cannot be excited through electric/magnetic coupling. By contrast, if the particles are rotated, a non-negligible component of the electric/magnetic field orthogonal to the symmetry plane of the SRRs/CSRRs appears, and both electric and magnetic couplings must be considered. Obviously, the optimum case to favor mixed coupling is the one where the slits are oriented in parallel to the line axis. Hence, the dependence of the frequency response of SRR- and CSRR-loaded lines on particle orientation is attributed to mixed coupling effects. To support this hypothesis, SRR- and CSRR-loaded lines (Figs. 3.2 and 3.3) subjected to mixed coupling are analyzed in terms of rotation, and the results can be seen in Figs. 3.4 and 3.5. As can be seen, essentially, the reflection coefficients change with rotation. When the structures are not identical seen from the ports ($\theta = 90°$), the reflection coefficients are different, the two ports can be topologically distinguished, and they cannot be freely interchanged.

In addition, the analysis considering a non-bianisotropic SRR (NB-SRR) [16] and a non-bianisotropic CSRR (NB-CSRR) is also performed. These particles, which are

[3]On the other hand, $|S_{11}| = |S_{22}|$ and $S_{12} = S_{21}$ as results from unitarity (lossless assumption) and reciprocity [15].

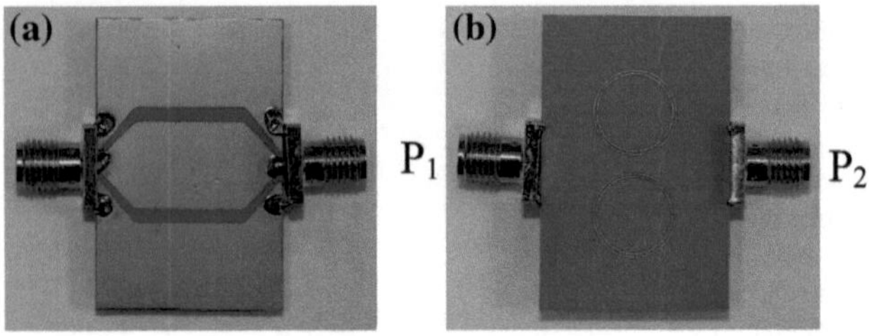

Fig. 3.2 Photograph of the **a** top and **b** bottom views of the considered SRR-loaded CPW with $\theta = 90°$. The substrate is *Rogers RO3010* with thickness $h = 1.27$ mm, relative permittivity $\epsilon_r = 11.2$, and loss tangent $\tan\delta = 0.0023$. Dimensions are: for the CPW, $W = 10.4$ mm, $G = 1.6$ mm (50-Ω line), and $l = 10.4$ mm; for the SRR, $c = d = 0.2$ mm and $r_{ext} = 5$ mm. Access lines to solder the connectors are included, so that the total physical length of the fabricated structure is longer than l

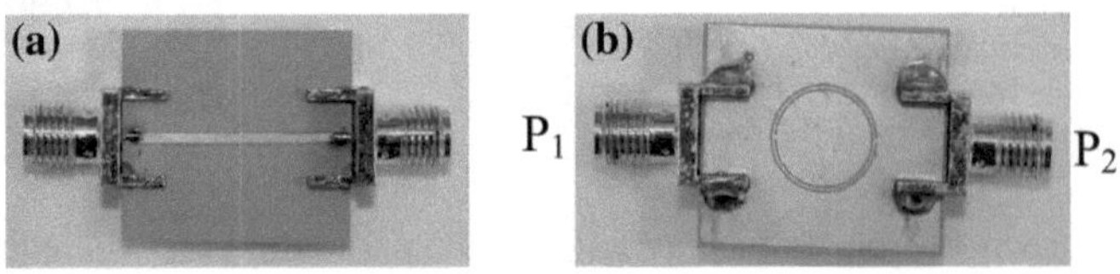

Fig. 3.3 Photograph of the **a** top and **b** bottom views of the considered CSRR-loaded microstrip with $\theta = 90°$. The substrate is *Rogers RO3010* with $h = 1.27$ mm, $\epsilon_r = 11.2$, and $\tan\delta = 0.0023$. Dimensions are: for the microstrip line, $W = 1.15$ mm (50-Ω line), and $l = 10.4$ mm; for the CSRR, $c = d = 0.2\ mm$ and $r_{ext} = 5$ mm. Access lines to solder the connectors are also included

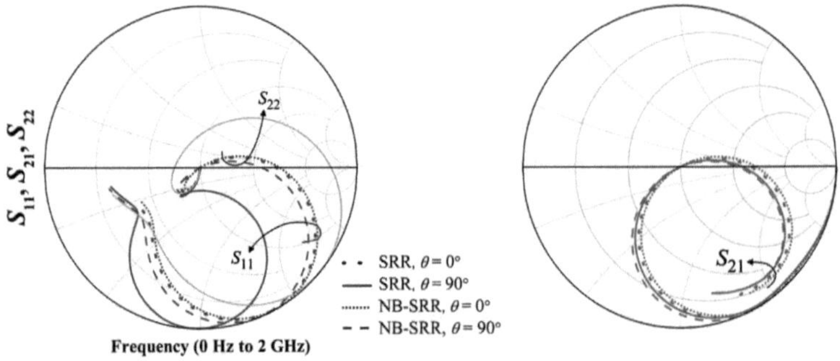

Fig. 3.4 Lossless frequency response in Smith Chart of a CPW loaded with a pair of SRRs or NB-SRRs for the two extreme rotation angles. The considered substrate and dimensions are those of Fig. 3.2. Reprinted with permission from [7]

not symmetric, do not exhibit cross-polarization since they have inversion symmetry with regard to their center (see Fig. 3.6). In other words, in contrast to resonators exhibiting a single symmetry plane (e.g. SRRs and CSRRs), which suffer from mixed coupling, resonators with inversion symmetry prevent from magnetoelectric

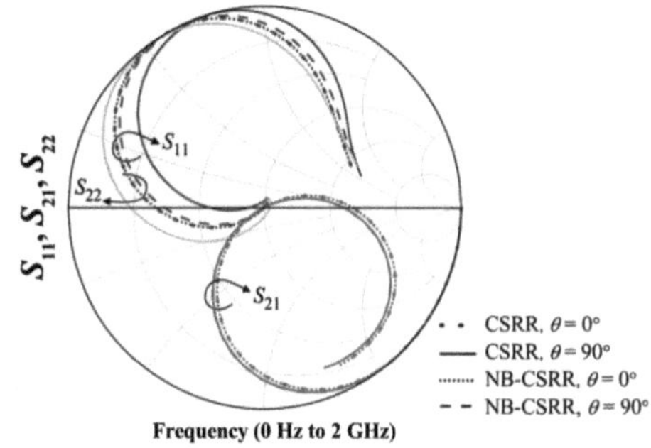

Fig. 3.5 Lossless frequency response in Smith Chart of a microstrip line loaded with a CSRR or NB-CSRR for the two extreme rotation angles. The considered substrate and dimensions are those of Fig. 3.3. Reprinted with permission from [7]

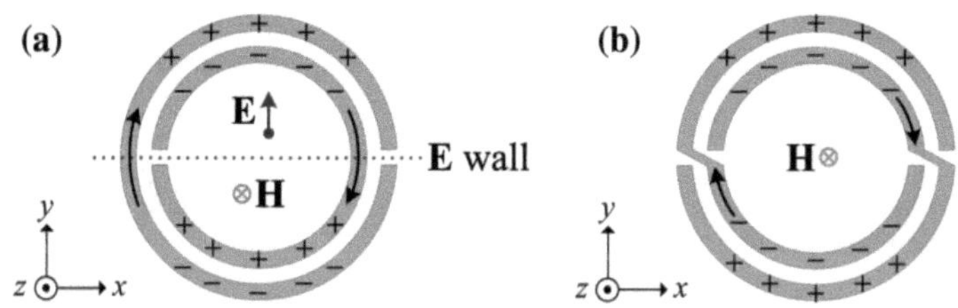

Fig. 3.6 Orientation of the polarization fields, and sketch of currents and charge distribution at the first resonance of **a** an SRR and **b** a NB-SRR. For the SRR, an electric dipole is induced along the y-direction. For the NB-SRR, there is no net electric dipole in the y-direction, and this particle cannot be excited by a time-varying electric field applied to this direction. By electromagnetic duality, it follows that the NB-CSRR cannot be driven by a magnetic field applied to the y-direction

coupling.[4] The results, plotted also in Figs. 3.4 and 3.5, show that the frequency response is almost invariant by rotating the NB-SRR/NB-CSRR, from which it can be drawn that mixed coupling alters the frequency response in SRR- and CSRR-loaded lines.

Therefore, unless the resonator slits are aligned along the orthogonal direction to the line axis, mixed magnetoelectric coupling between the host line and the resonators must be considered to accurately describe the lines with arbitrarily oriented SRRs or CSRRs. To this end, the equivalent circuit models of SRR-loaded CPWs and CSRR-loaded microstrip lines, reproduced in Figs. 3.7a and 3.8a, are updated including mixed coupling, as represented in Figs. 3.7b and 3.8b. A capacitance and a mutual inductance are introduced in the SRR- and CSRR-based circuits to model the line-to-resonator capacitive and inductive coupling, respectively (see Appendix B). It is observed that the magnetic coupling provides the necessary asymmetry in the

[4]ELC and MLC resonators exhibit inversion symmetry. However, in ELC- and MLC-loaded differential microstrip lines, where the fields are non-uniform by far, mixed coupling is present (Sects. 4.2.3) [9].

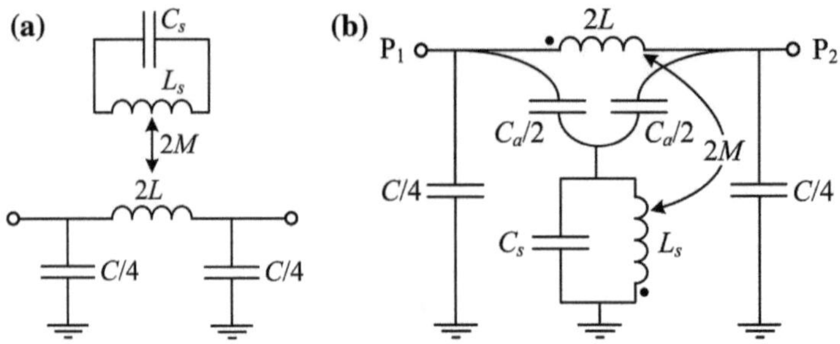

Fig. 3.7 Unit-cell equivalent circuit model of a CPW loaded with pairs of SRRs, applying the magnetic wall concept (open circuit). **a** Model without mixed coupling which is strictly valid only for $\theta = 0°$ [10–12]. **b** Proposed circuit model valid for any arbitrary SRR angular orientation [7]. It is assumed that $C_a = f(\theta)$, where $C_a(\theta = 0°) = 0$

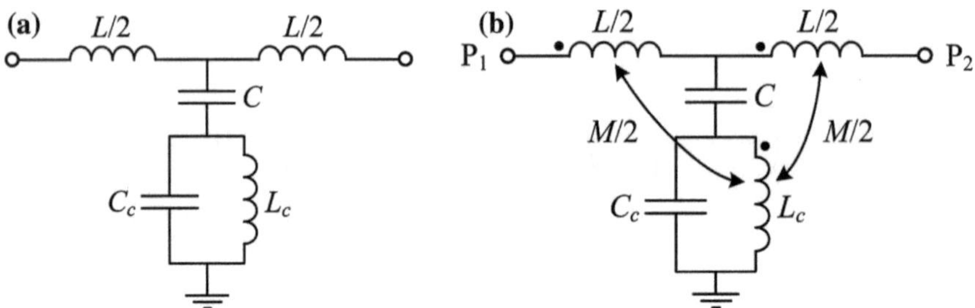

Fig. 3.8 Unit-cell equivalent circuit model of a microstrip line loaded with CSRRs. **a** Model without mixed coupling which is strictly valid only for $\theta = 0°$ [6]. **b** Proposed circuit model valid for any arbitrary CSRR angular orientation [7]. It is assumed that $M = f(\theta)$, where $M(\theta = 0°) = 0$

circuit models to explain the lack of symmetry with regard to the ports ($S_{11} \neq S_{22}$).[5] However, in the SRR-based circuit, it is important to realize that this asymmetry is not manifested if electric coupling is not present, and the electric coupling is indeed responsible for the asymmetry. Note that because of the asymmetric topology (in both the structure and the circuit), it is mandatory to identify the two ports with no ambiguity, as it is by P_1 and P_2. It is also of relevance that, like the models without mixed coupling, the resulting equivalent circuit models of SRR/CSRR-loaded lines in the presence of magnetoelectric coupling satisfy the duality property.

The presented equivalent circuits are validated by comparing circuit simulations with electromagnetic simulations and measurements in Figs. 3.9 and 3.10, where good agreement is obtained. It can be clearly appreciated that the phases of the two reflection coefficients are different from each other, as expected.

Finally, it should be highlighted that the proposed equivalent circuits are considerably complex, and essentially such circuits allow for the accurate modeling of

[5]The dot convention in the inductive coupling is used, where the current entering a dot-marked terminal in one inductor produces a positive voltage referenced at the dotted terminal of the other inductor [17]. Thereby there is no ambiguity in the sign of the induced voltage (Appendix B).

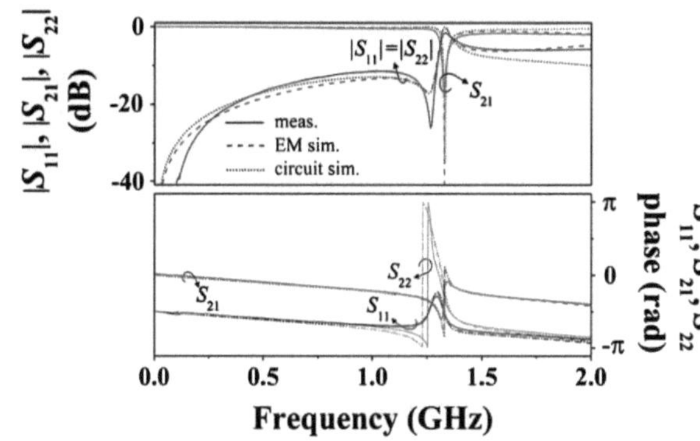

Fig. 3.9 Frequency response of the SRR-loaded CPW in Fig. 3.2. Losses in the simulations are excluded. The circuit parameters are extracted as follows. (*i*) For $\theta = 0°$, the parameters in the transformed circuit of Fig. 2.12b are obtained according to the procedure in [13]; $L' = 2.5$ nH, $C = 6$ pF, $L'_s = 0.4$ nH, $C'_s = 38.8$ pF, and $L = 1.45$ nH. (*ii*) The parameters of the circular SRRs are estimated analytically from the model in [1, 2]; $L_s = 27.8$ nH and $C_s = 0.6$ pF. (*iii*) For $\theta = 90°$, L_s, C_s, M, and C_a are adjusted by curve fitting the circuit simulation to the electromagnetic simulation, such that the resulting L_s and C_s are similar to those obtained analytically; $L = 1.5$ nH, $C = 6.1$ pF, $L_s = 28.2$ nH, $C_s = 0.43$ pF, $M = 1.35$ nH, and $C_a = 0.08$ pF. For simplicity, the magnetic wall concept is applied to the network, but not to the ports. In the measurements, the access lines are de-embedded so that the length of the structure becomes l. Reprinted with permission from [7]

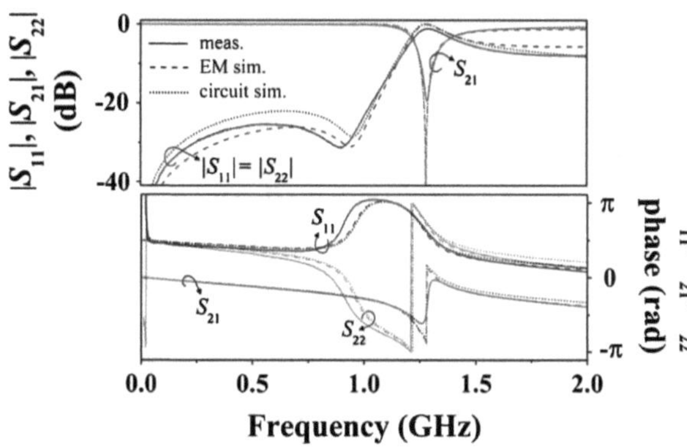

Fig. 3.10 Frequency response of the CSRR-loaded microstrip line in Fig. 3.3. Losses in the simulations are excluded. The circuit parameters are extracted as follows. (*i*) For $\theta = 0°$, the parameters of the circuit in Fig. 3.8a are obtained according to the procedure explained in [14]; $L = 7.87$ nH, $C = 1.62$ pF, $L_c = 2.94$ nH, and $C_c = 3.9$ pF. (*ii*) For $\theta = 90°$, M is extracted by curve fitting, and a slight optimization of the other parameters is carried out; $L = 7$ nH, $C = 1.7$ pF, $L_c = 2.74$ nH, $C_c = 3.95$ pF, and $M = 0.27$ nH. In the measurements, the reference planes are shifted. Reprinted with permission from [7]

the phase response of the reflection coefficients.[6] Therefore, unless accurate phase response modeling is required, rather than being useful for design purposes, the circuits including mixed coupling (*i*) expand the understanding of the associated line-to-resonator coupling phenomenology, (*ii*) point out the relevance of particle orientation on its excitation, and (*iii*) indicate that the parameter extraction procedures may fail when the symmetry plane of the resonators is not aligned orthogonally with the line.

3.2 Inter-Unit-Cell Inter-Resonator Coupling

The previous section was devoted to the line-to-resonator coupling. Additionally, in a transmission line loaded with an array of resonators, the study of the coupling between nearby resonators is also meaningful. Likewise, the coupling between two resonant elements depends on their relative orientation, and may be electric, magnetic, or mixed [18–22]. The aim of this section is to consider inter-resonator coupling between resonators of adjacent unit cells.[7] As in Sect. 2.4, the coupling is assumed to be restricted to the first-neighbor interaction[8] and to the dominant coupling mechanism (i.e. mixed coupling is not considered). As will be shown, inter-resonator coupling enhances the stopband bandwidth, and this is useful for the design of bandstop filters. For instance, in Sect. 5.3.1, inter-resonator coupling helps to broaden the stopband of the common mode in CSRR-loaded differential microstrip lines.

3.2.1 *Coplanar Waveguides Loaded with Pairs of SRRs and CSRR-Loaded Microstrip Lines*

Transmission lines periodically loaded with SRRs or CSRRs inhibit wave propagation in the vicinity of the resonator fundamental resonance, and support forward waves outside this forbidden band (Sect. 2.2.4). It is also known that chains of coupled SRRs (and other magnetic resonators) or CSRRs (and other electric resonators)

[6]The transmission zero frequency is additionally affected by both electric and magnetic couplings. However, for the considered (typical) structures, the non-dominant coupling mechanism has a small impact on that frequency.

[7]This section deals with coupling between resonators of different cells, here termed *inter-unit-cell coupling*. However, when the unit cell of resonator-loaded transmission lines is composed of more than one resonator, inter-resonator coupling within the unit cell may also exist, and may be termed *intra-unit-cell coupling*. Intra-unit-cell coupling in cells composed of pairs of resonators is analyzed in Sects. 4.4.1 and 4.4.2.

[8]Coupling between each resonator and a number of neighbors may be significant. The decay of the coupling coefficient, that varies with inter-resonator separation, determines the number of couplings that should be taken into account. However, it is difficult to model and design structures if many couplings are involved.

arranged on the same plane support backward waves, and such waves have been called magneto-inductive (MIWs) and electro-inductive waves (EIWs), respectively (Sect. 2.4). Interestingly, coupling manifestations between MIWs/EIWs and electromagnetic waves (e.g. propagating through an array of SRRs/CSRRs and a transmission line, respectively) may arise.

In the literature, the dispersion characteristics of transmission lines loaded with SRRs (considering coupled and decoupled SRRs) and metallic rods were analyzed by Syms et al. [23]. It was shown that MIWs and electromagnetic waves may interact leading to a stopband. Some years later, Syms and Solymar carried out a theoretical analysis of SRR-loaded lines with coupled resonators (without the rods), and in a region of the stopband they found the presence of a pair of complex waves characterized by complex propagation constants [24].[9] Shortly after, a similar study on the coupling between MIWs and TEM (transverse electromagnetic) waves was reported in [25].

The complex waves found in [24] are referred to as *complex modes* in the literature, which have been known for a long time, and have been found in several structures [26–30]. Despite the assumption of the absence of losses, complex modes have complex propagation constants and appear as conjugate pairs in reciprocal structures ($\gamma = \alpha \pm j\beta$, where α is the attenuation constant and β is the phase constant). Therefore, complex modes exhibit the same α, and the same β but with opposite sign. Since the power contributions of the individual modes are in opposite directions to each other, complex modes do not generally carry net power.[10]

SRR- and CSRR-loaded lines with coupled resonators are studied by Naqui et al. in [32, 33] and [34], respectively. The particular topologies under study are depicted in Fig. 3.11. The topology of the considered SRR-loaded line (unit cell) consists of a CPW loaded with pairs of SRRs etched on the back side of the substrate. The considered CSRR-loaded line is a microstrip line loaded with an array of CSRRs etched in the ground plane. To achieve significant inter-resonator coupling, closely-spaced and rectangular-shaped resonators (the long side being oriented along the inter-resonator interfaces) are designed. It is important to highlight that the symmetry plane of the resonators (along the slits) is oriented orthogonally to the line axis. This orientation guarantees that the line is capable of exciting the SRR/CSRR fundamental resonance through only the magnetic/electric coupling. With different orientations, line-to-resonator mixed coupling is required for an accurate description of the structures, as discussed in Sect. 3.1.1 (obviously, mixed coupling between the line and the resonators makes the present analysis much more cumbersome).

To study the inter-resonator interaction, resonator-to-resonator coupling is included (assuming the nearest-neighbor interaction approximation) into the equivalent circuit models of SRR/CSRR-loaded lines, namely, those depicted in Figs. 3.7a

[9] According to [24], in an SRR-loaded line with coupled resonators, the region of complex modes leads to a complex effective permeability (under lossless conditions).

[10] Under some circumstances, the individual complex modes may be selectively excited [31].

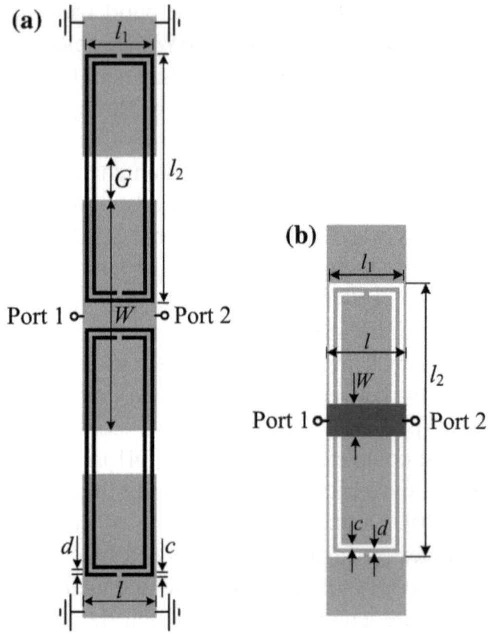

Fig. 3.11 Unit cell of **a** a CPW transmission line loaded with a pair of rectangular-shaped SRRs and **b** a microstrip transmission line loaded with a rectangular-shaped CSRR etched in the ground plane. The considered substrate is *Rogers RO3010* with thickness $h = 1.27$ mm, relative permittivity $\epsilon_r = 11.2$, and loss tangent $\tan\delta = 0.0023$. Dimensions are: the unit cell size is $l = 3$ mm; for the CPW: $W = 9.1$ mm and $G = 1.7$ mm (50-Ω line); for the microstrip line, $W = 1.17$ mm (50-Ω line); for the SRR/CSRR: $l_1 = 2.8$ mm, $l_2 = 9.8$ mm, and $c = d = 0.15$ mm

and 3.8a.[11] The resulting unit-cell equivalent circuit models are schematically represented in Figs. 3.12 and 3.13. It should bear in mind that the dominant coupling mechanism is modeled only, that is, the inter-resonator coupling is supposed to be dominantly magnetic and electric in SRRs and CSRRs, respectively (the mutual inductance between SRRs of adjacent cells being negative). Even though losses can be modeled by including resistors in the equivalent circuits, losses are neglected, and this is an important assumption when the dispersion relation (in the presence of complex modes) is derived. It is also worth pointing out that when the SRRs of the same unit cell are close together, there can be magnetic coupling between them of positive sign. However, this intra-unit-cell coupling is not considered because its only effect on the frequency response is a decrease in the resonance frequency.

As is apparent, the lumped-element equivalent circuit models are four-port circuits. Qualitatively, the four-port unit-cell circuit of SRR-loaded lines with coupled resonators (Fig. 3.12c) is composed of a conventional two-port transmission line-type circuit magnetically coupled to a secondary propagating circuit that supports the propagation of MIWs. Analogously, the unit-cell circuit of CSRR-loaded lines when the resonators are coupled (Fig. 3.13) can be regarded as a transmission-line circuit electrically coupled to a circuit that is able to support EIWs. As multiconductor theory predicts, both unit cells can support two modes [35]. While a conventional transmission line supports the propagation of forward waves (Sect. 2.2.1), an array

[11] Although inter-resonator coupling beyond the first neighbor may be significant using rectangular-shaped resonators, this approximation suffices to analyze the effects of inter-unit-cell coupling.

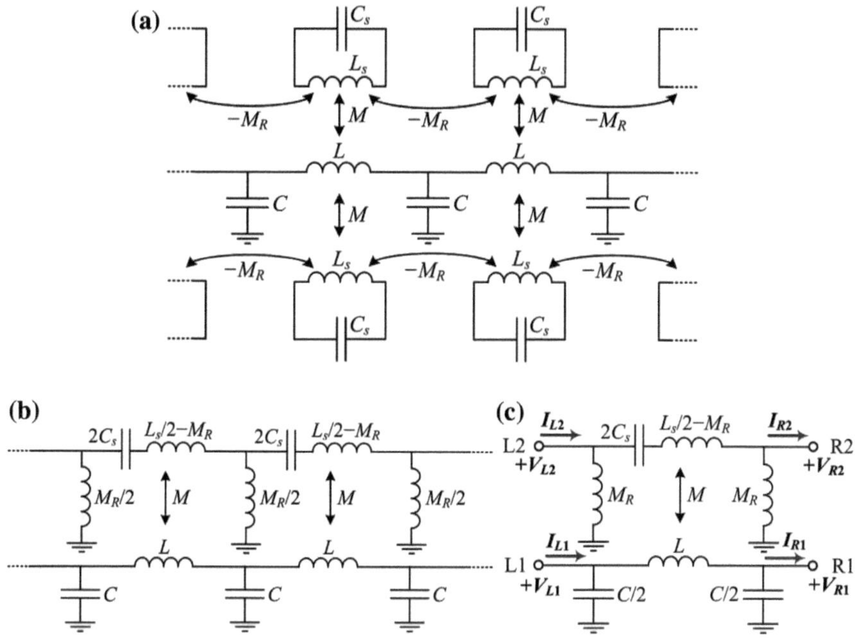

Fig. 3.12 Equivalent circuit model of a periodic SRR-loaded CPW composed of an array of unit cells like the one shown in Fig. 3.11, including magnetic coupling between resonators of adjacent cells. **a** Model with mutual inductance between resonators. **b** Transformed model (see Appendix B.2). **c** Four-port unit cell model indicating the port voltages and currents relative to the calculation of the *ABCD* matrix. The elements L and C are the per-section line inductance and capacitance, respectively, the SRR is described by the capacitance C_s and the inductance L_s, M stands for the mutual inductance between the line and the SRRs, and inter-resonator coupling is accounted for through the mutual inductance M_R

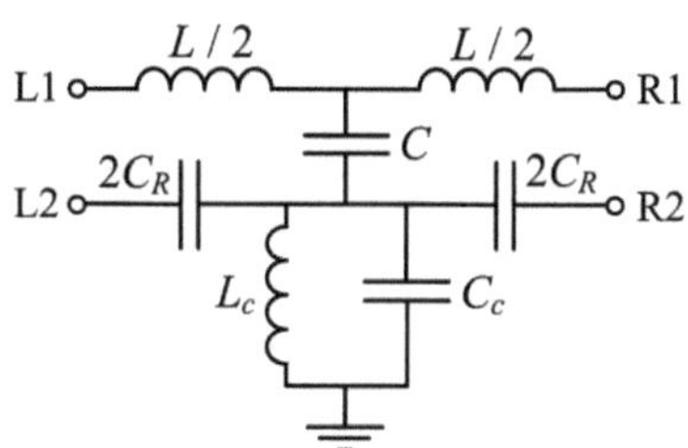

Fig. 3.13 Four-port unit-cell equivalent circuit model of a periodic CSRR-loaded microstrip line composed of an array of unit cells like the one shown in Fig. 3.11, including electric coupling between resonators of adjacent cells. The parameter L models the microstrip line inductance, C accounts for the line capacitance as well as the electric coupling between the line and the CSRR, modeled by L_c and C_c, and C_R describes the coupling between adjacent resonators

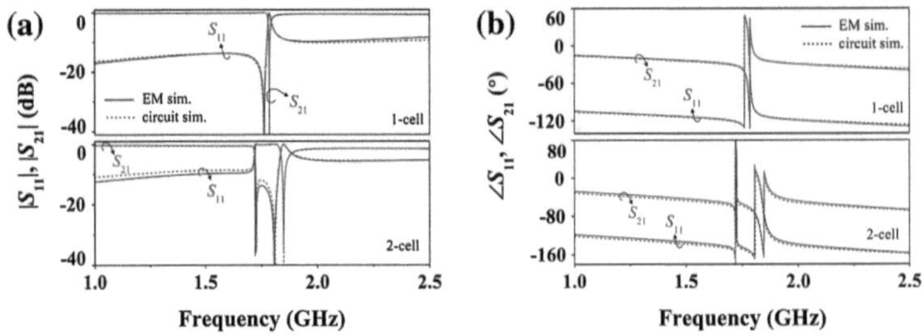

Fig. 3.14 **a** Magnitude and **b** phase of the lossless transmission (S_{21}) and reflection (S_{11}) coefficients for a unit cell and for two cascaded unit cells of the SRR-loaded CPW in Fig. 3.11. The circuit parameter extraction is as follows. (*i*) The elements of the transformed circuit model corresponding to an isolated unit cell ($M_R = 0$) are extracted using the method reported in [13]. (*ii*) Next, L_s is estimated as the self-inductance of an isolated single split ring with the same average radius and ring width as the considered SRRs using the technique reported in [36]. (*iii*) By using the estimated L_s, the circuit elements of an isolated unit cell have been obtained from the transformation equations; $L = 1.01$ nH, $C = 1.40$ pF, $L_s = 17.66$ nH, $C_s = 0.45$ pF, and $M = 0.72$ nH. (*iv*) Finally, $M_R = 1.17$ nH is inferred by curve fitting the frequency response in a 2-cell structure. Reprinted with permission from [33]

of coupled resonators in coplanar arrangement supports either MIWs or EIWs of backward nature (Sect. 2.4.1 and 2.4.2, respectively). In consequence, forward and backward waves are expected to coexist at some frequency band.

It is important to realize that, in a truncated periodic structure, the resonators of the input/output cells are not externally fed. Only the transmission line is driven by port terminals, and then MIWs/EIWs are excited from the electromagnetic waves generated by the line that illuminate the resonators. Therefore, when the periodic structure is truncated, the ports modeling the inter-resonator coupling are open-ended, and a simple two-port frequency response is computed.

With a view to validating the circuit models, the transmission and reflection coefficients corresponding to the circuit and electromagnetic simulations are compared. For one-cell (where there is no inter-resonator coupling) and two-cell (where there is unilateral inter-resonator coupling only, i.e. nearest-neighbor interaction) structures, the accordance of the circuit responses with those obtained from electromagnetic simulations is remarkable (Figs. 3.14 and 3.15). For higher-order structures, where the nearest-neighbor approximation plays an active role, the accuracy of the circuit model drops to some extent. This is apparent in Figs. 3.16a and 3.17a where the frequency response of order-9 structures is shown. It should be also noticed that the central frequency of the stopbands shifts upward as the number of cells increases, this effect being produced by the inter-resonator coupling. In this regard, note that the stopband inferred from the electromagnetic simulation is even slightly more shifted than the one predicted by the circuit simulation. It has been found (by including additional couplings between non-adjacent cells) that this discrepancy is due to the first-neighbor approximation. To support this statement in an alternative way, an increase in the inter-resonator distance reduces the impact on the frequency

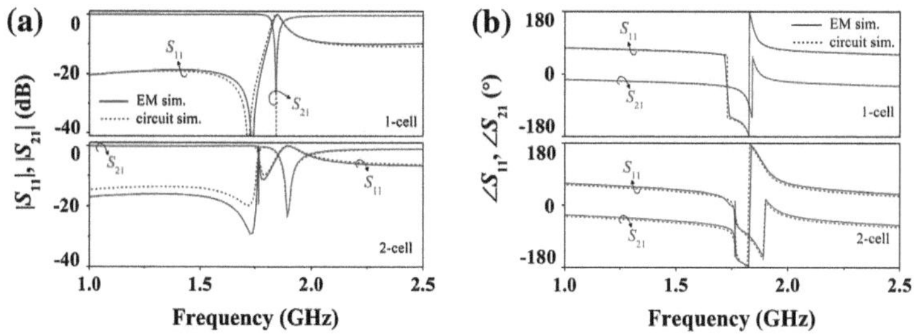

Fig. 3.15 **a** Magnitude and **b** phase of the lossless transmission and reflection coefficients for a unit cell and for two cascaded unit cells of the CSRR-loaded microstrip line in Fig. 3.11. The circuit parameters of an isolated unit cell ($C_R = 0$) have been extracted according to the method reported in [14]; $L = 3.05$ nH, $C = 0.55$ pF, $L_c = 2.03$ nH, and $C_c = 3.12$ pF. Afterward, for the two-order structure, the value of $C_R = 0.21$ pF and the new value of $C_c = 2.98$ pF (the CSRR capacitance is reduced by the presence of neighboring CSRRs) have been inferred by direct fitting to the electromagnetic simulation. Reprinted with permission from [34]

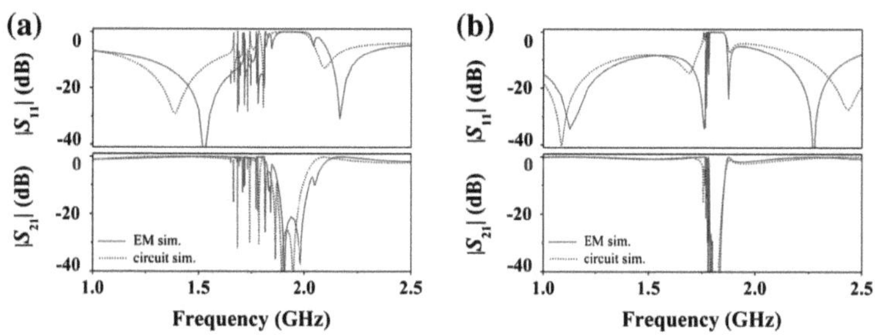

Fig. 3.16 Lossless electromagnetic and circuit simulations of the transmission and reflection coefficients (magnitude) for nine cascaded elemental cells of the SRR-loaded CPW in Fig. 3.11 for **a** $l = 3$ mm and **b** $l = 4.8$ mm (the inter-resonator spacing has been increased from 0.2 to 2 mm). The circuit parameters are those indicated in the caption of Fig. 3.14, with the exception of $M_R = 0.22$ nH in **b**. Furthermore, in **b**, a CPW section of 1.8 mm length has been cascaded between the ports R1 and L1 of contiguous cells in the circuit model. Reprinted with permission from [33]

response shift caused by such an approximation. As Figs. 3.16b and 3.17b confirm, the agreement between the electromagnetic and circuit simulations is better, and the stopbands occur at lower frequency regions. Nevertheless, the first-neighbor approximation suffices for the purpose of the present study.

It is important to point out that inter-resonator coupling splits the fundamental resonance frequency, so that the number of transmission zero frequencies equals the number of resonators (without coupling there would be a single resonance frequency). As is well known, the stronger the coupling, the stronger the frequency splitting. Accordingly, when multiple resonators are coupled, the stopband bandwidth broadens with inter-resonator coupling, though this is at the expense of reducing the stopband attenuation.

The dispersion characteristics of the considered resonator-loaded lines are also obtained from Bloch-mode theory applied to the four-port networks. To derive the

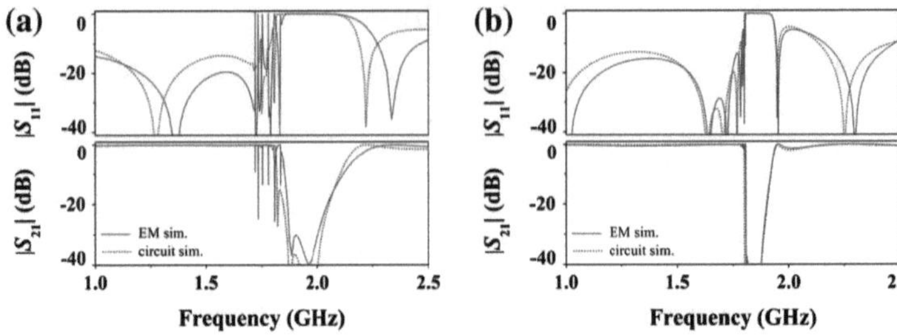

Fig. 3.17 Lossless electromagnetic and circuit simulations of the transmission and reflection coefficients (magnitude) for nine cascaded elemental cells of the CSRR-loaded microstrip line in Fig. 3.11 for **a** $l = 3$ mm and **b** $l = 4.8$ mm. The circuit parameters are those indicated in the caption of Fig. 3.15, with the exception of: $C_c = 2.84$ pF (readjusted to account for bilateral contiguous couplings) in **a**, and $C_R = 0.017$ pF and $C_c = 3.2$ pF in **b**. Moreover, in **b**, a microstrip line section of 1.8 mm length has been cascaded between the ports R1 and L1 of contiguous cells in the circuit model. Reprinted with permission from [34]

dispersion relation, a complex propagation constant, $\gamma = \alpha + j\beta$, is assumed. As is shown in Fig. 3.12c, the voltages and currents at the ports ($i = 1, 2$) of the left-hand side of the unit cell are denoted by V_{Li} and I_{Li}, whereas V_{Ri} and I_{Ri} are the variables at the right-hand side ports. The variables at both sides of the two networks are linked through a generalized order-4 transmission (*ABCD*) matrix, as is typically performed in two-port networks (Appendix A). Specifically, the currents and voltages at both sides of the networks are linked by

$$\begin{pmatrix} \boldsymbol{V_L} \\ \boldsymbol{I_L} \end{pmatrix} = \begin{pmatrix} \boldsymbol{A} & \boldsymbol{B} \\ \boldsymbol{C} & \boldsymbol{D} \end{pmatrix} \begin{pmatrix} \boldsymbol{V_R} \\ \boldsymbol{I_R} \end{pmatrix}, \tag{3.1}$$

where $\boldsymbol{V_L}$, $\boldsymbol{I_L}$, $\boldsymbol{V_R}$, and $\boldsymbol{I_R}$ are column vectors composed of the pair of port variables, and $\boldsymbol{A}$, $\boldsymbol{B}$, $\boldsymbol{C}$, and $\boldsymbol{D}$ are order-2 matrices. The dispersion relation is obtained from the eigenmodes of the system (3.1), which are inferred by forcing a given propagation factor along the unit block, and setting the determinant of the resulting matrix to be null, that is:

$$\det \begin{pmatrix} \boldsymbol{A} - e^{\gamma l} \cdot \boldsymbol{I} & \boldsymbol{B} \\ \boldsymbol{C} & \boldsymbol{D} - e^{\gamma l} \cdot \boldsymbol{I} \end{pmatrix} = 0, \tag{3.2}$$

where $\boldsymbol{I}$ is the identity matrix, the propagation factor $e^{\gamma l}$ is the eigenvalue, and l is the unit cell length. For reciprocal, lossless, and symmetric networks, the two[12] modal solutions for the complex propagation constant can be easily found by [37]

$$\det\left(\boldsymbol{A} - \cosh(\gamma l) \cdot \boldsymbol{I}\right) = 0, \tag{3.3}$$

[12]There are actually four mathematical solutions corresponding to the complete set of forward- and backward-traveling waves [15]. For convention, only the solutions associated to positive group velocity are considered, which represent power flow in the reference positive direction.

which gives

$$\cosh(\gamma l) = \frac{1}{2}\left(A_{11} + A_{22} \pm \sqrt{(A_{11} - A_{22})^2 + 4A_{12}A_{21}}\right), \tag{3.4}$$

where the elements of the $\boldsymbol{A}$ matrix obtained from the networks of Figs. 3.12c and 3.13 are, respectively,

$$\boldsymbol{A} = \boldsymbol{D}^t = \begin{pmatrix} 1 - \frac{LC}{2}\omega^2 & \frac{M}{M_R} \\ -\frac{MC}{2}\omega^2 & \frac{1}{2M_R}\left(L_s - \frac{1}{C_s\omega^2}\right) \end{pmatrix}, \tag{3.5}$$

and

$$\boldsymbol{A} = \boldsymbol{D}^t = \begin{pmatrix} 1 - \frac{LC}{2}\omega^2 & \frac{LC}{2}\omega^2 \\ -\frac{C}{2C_R} & 1 + \frac{C}{2C_R} + \frac{C_c}{2C_R}(1 - \frac{1}{L_cC_c\omega^2}) \end{pmatrix}. \tag{3.6}$$

Since the considered circuits are lossless, the elements of $\boldsymbol{A}$ (A_{ij}) are real numbers. Hence, if the radicand in (3.4) is positive, the propagation constant is either purely real ($\alpha \neq 0$ and $\beta = 0$) or purely imaginary ($\alpha = 0$ and $\beta \neq 0$), corresponding to evanescent or propagating modes, respectively. However, if the radicand in (3.4) is negative, the two solutions are of the form $\gamma = \alpha \pm j\beta$, corresponding to complex modes. The frequency band that supports complex modes is thus obtained by forcing the radicand in (3.4) to be negative, i.e.

$$(A_{11} - A_{22})^2 + 4A_{12}A_{21} < 0. \tag{3.7}$$

A necessary (although not sufficient) condition to have complex modes is an opposite sign for the reactances of the two series branches.

The results of the analytical dispersion relations, plotted in Figs. 3.18 and 3.19 show that, in the first allowed band, there is a region with bi-valued propagation constant (i.e. with multimode propagation): one mode is forward (where the propagation constant is purely imaginary and positive) corresponding to transmission-line propagation, and the other one is backward (the propagation constant being purely imaginary and negative) related to MIWs (in SRR-based lines) or EIWs (in CSRR-based lines). Then, a region with a pair of complex modes appears ($\gamma = \alpha \pm j\beta$),[13] followed by a region of evanescent waves (where the propagation constants are purely real). Finally, a forward-wave transmission band emerges again. The results obtained from Bloch-mode theory are therefore in good agreement with those reported by Syms and Solymar [24]. Since complex modes do not carry net power, the frequency band supporting such modes is a rejection band. Hence, complex modes contribute to stopband

[13]The dispersion relation of a circuit that is formally the same as that corresponding to CSRR-loaded lines is obtained from a four-port network impedance matrix by Tsai and Wu [38], but the appearance of complex modes is not reported therein.

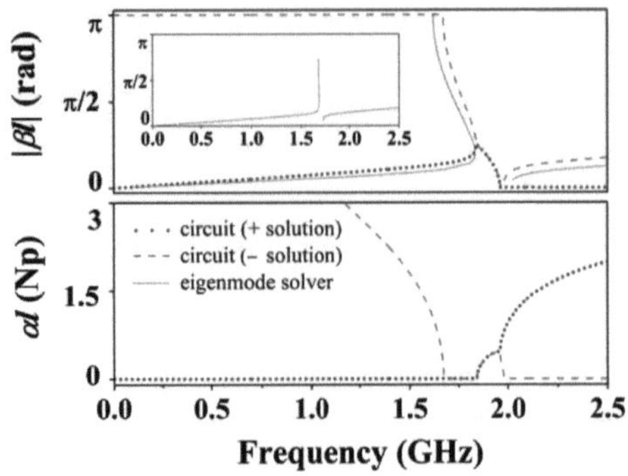

Fig. 3.18 Dispersion relation for the SRR-loaded CPW in Fig. 3.11 obtained from its equivalent circuit model of Fig. 3.12c and from an eigenmode solver. The circuit parameters are those indicated in the caption of Fig. 3.14. The dispersion relation with $l = 4.8$ mm is plotted in the inset. Reprinted with permission from [33]

broadening in SRR- and CSRR-loaded lines with coupled resonators.[14] It is apparent that the stopband characteristics in the complex wave region can be interpreted as a result of the destructive interference of antiparallel forward transmission-line propagation and backward MIWs/EIWs. Note that the behavior of SRR- and CSRR-loaded lines with coupled resonators is very similar to one another. The difference is the nature of the backward waves, which belongs to MIWs or EIWs in SRR- and CSRR-loaded lines, respectively.

The modal solutions of an infinite periodic structure composed of a cascade connection of the corresponding SRR- and CSRR-based unit cells is also obtained by means of the full-wave eigenmode solver in *Computer Simulation Technology (CST) Microwave Studio*. This solver infers $|\beta l|$, and no information is provided about the attenuation constant α or the complex mode region. Nevertheless, as can be observed in Figs. 3.18 and 3.19, there is good agreement with the theoretical dispersion curves predicted by the circuit models. For comparison purposes, structures with wider inter-resonator separation providing loosely coupled resonant elements are in addition simulated. The dispersion diagrams, included in the insets of Figs. 3.18 and 3.19, reveal that the resulting stopband bandwidth is significantly narrower. It is thus drawn that most of the stopband in the considered structures is related to the presence of complex, rather than evanescent, modes. Therefore, as long as inter-resonator coupling is relatively tight, complex modes are expected to be the dominant mechanism of signal rejection.

Finally, to experimentally confirm the effects of inter-resonator coupling on bandwidth enhancement, the frequency response of the four order-9 structures under study

[14] Analogously, in [34] it is also drawn that in composite right-/left-handed (CRLH) structures based on transmission lines loaded with CSRRs and series gaps (Sect. 2.2.3), the left-handed passband broadens as inter-resonator coupling strengthens. The reason is a constructive interference between EIWs and the backward-wave propagation of the CRLH structure. Prior circuit modeling of these structures without applying Bloch-mode theory was presented by Gil et al. [39].

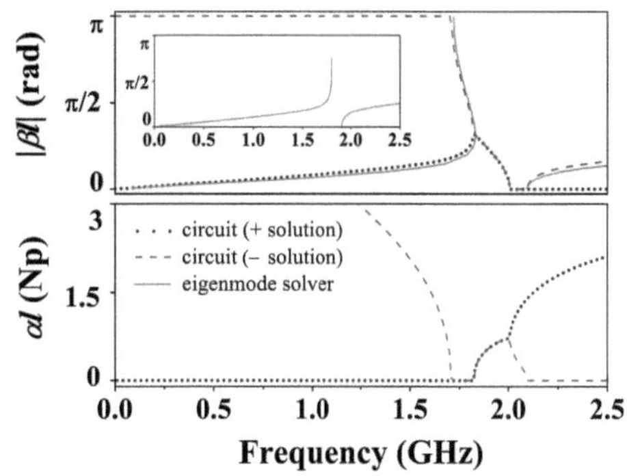

Fig. 3.19 Dispersion relation for the CSRR-loaded microstrip line in Fig. 3.11 obtained from its equivalent circuit model of Fig. 3.13 and from an eigenmode solver. The circuit parameters are those indicated in the caption of Fig. 3.15, except $C_c = 2.84$ pF (readjusted to account for two first-neighbor couplings). The dispersion relation with $l = 4.8$ mm is plotted in the inset. Reprinted with permission from [34]

is measured (see Figs. 3.20 and 3.21); a pair with nearby resonators (providing a strong coupling), and another pair with distant resonators (achieving a weak coupling). For the SRR-loaded CPWs, the measured fractional stopband bandwidth at −20 dB is 5.2 and 2.4 % for close and distant resonators, respectively. For the CSRR-loaded microstrip lines, these values are 9.4 and 6.2 %. The measured stopbands are also in accordance with those obtained from the dispersion relation.

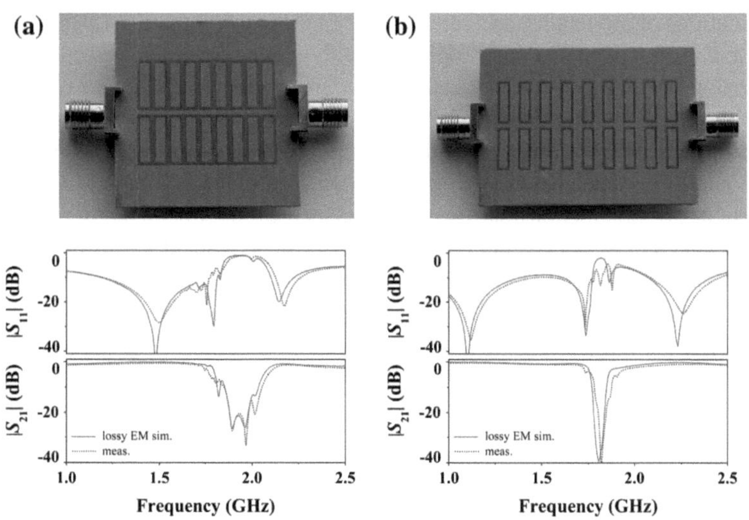

Fig. 3.20 Photograph of the bottom face of the fabricated order-9 SRR-loaded CPWs composed of the unit cell in Fig. 3.11, and lossy electromagnetic simulations and measurements. **a** $l = 3$ mm and **b** $l = 4.8$ mm. Reprinted with permission from [33]

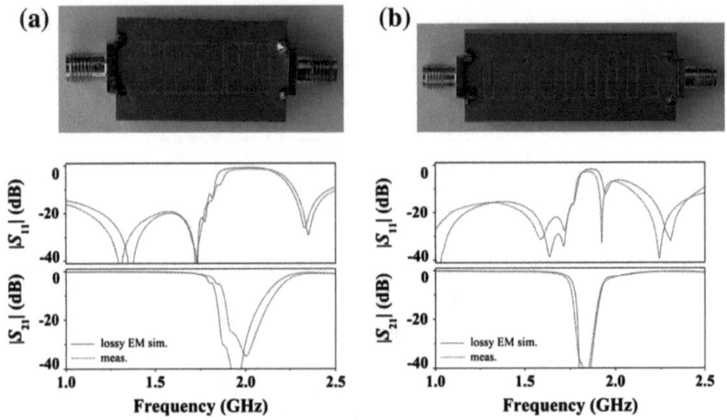

Fig. 3.21 Photograph of the bottom face of the fabricated order-9 CSRR-loaded lines composed of the unit cell in Fig. 3.11, and lossy electromagnetic simulations and measurements. **a** $l = 3$ mm and **b** $l = 4.8$ mm. Reprinted with permission from [34]

3.3 Limits on the Synthesis of Electrically Small Resonators

In microwave engineering, the *synthesis* of resonators concerns the design of a particular physical topology providing a certain inductance and capacitance, usually prescribed by design specifications (e.g. to widen the rejection bandwidth of a band-stop filter). Frequently, practical limitations exist when synthesizing resonators, that is, for a particular resonance frequency, it is not possible to achieve all the infinite combinations of inductance and capacitance. In order to estimate the inductance and capacitance of a given resonator topology, analytical models can be used. For example, analytical values of the inductance and capacitance of SRRs and CSRRs have been reported in the literature [1, 2, 6]. However, these analytical expressions do not give information about the electrical length of the resonators, and this is indeed of paramount importance for modeling the resonant particles as lumped *LC* circuits.

Particularly, this section analyzes the limits on the implementable *LC* values using stepped-impedance shunt-stubs (SISSs). Since these resonators are implemented through microstrip lines, and there are available well-established expressions for microstrip technology, the electrical size of the resonator may be determined straightforwardly. This analysis highlights the well-known practical limitations when synthesizing resonators, and is useful for the design of microwave components based on SISSs. For instance, the reported results have been applied to the automatic synthesis of SISSs through aggressive space mapping (ASM) [40, 41].

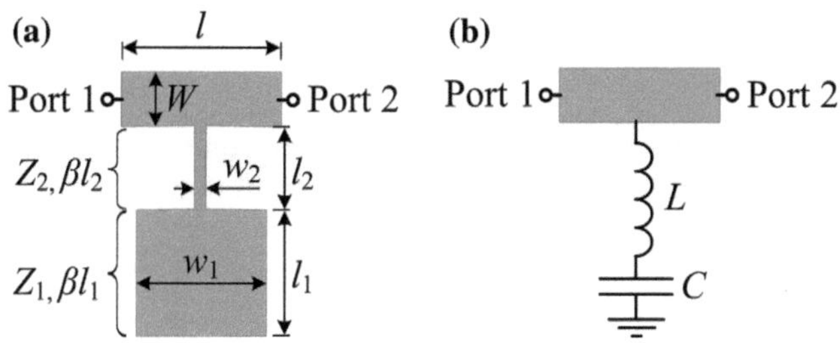

Fig. 3.22 **a** Topology of a SISS loading a microstrip transmission line ($Z_2 >> Z_1$), and **b** equivalent circuit model of the SISS

3.3.1 Microstrip Stepped-Impedance Shunt-Stubs (SISSs)

There are many different topologies and feedings of stepped-impedance resonators (SIRs), originally proposed by Makimoto and Yamashita [42]. In Sect. 2.3.3 we introduced the folded SIR (FSIR) which can be excited electrically and/or magnetically. In this section, by contrast, the focus is on shunt-connected SIRs which are excited by port terminals, as can be seen in Fig. 3.22a. Such SIRs can be viewed as stubs exhibiting a high-impedance section in contact with the host line, cascaded to an open-ended low-impedance section. It is for this reason that these resonators are also referred to as stepped-impedance shunt-stubs (SISS) [43, 44].

Naqui et al. study in [45] the limits on the implementation of shunt-connected series *LC* resonators by means of SISSs in microstrip technology, as illustrated in Fig. 3.22b. The aim of the work is to propose a procedure to determine the implementable inductance and capacitance, so that the response of the SISS accurately describes the response of the ideal lumped *LC* resonator up to a given frequency (obviously beyond resonance). The implementable values are given as a function of the frequency range of interest, the accuracy, the substrate parameters, and the fabrication technology. The higher the frequency range and the higher the accuracy, the smaller the implementable values. The implementable values are plotted in an *LC* plane for smart visualization.

The topology of a typical SISS, the relevant dimensions (lengths and widths of the transmission line sections) and the electrical parameters (characteristic impedances and electrical lengths) are depicted in Fig. 3.22a. The distributed admittance of the SISS (seen from the host line) is given by[15]

$$Y_{\mathrm{SISS}} = -j\frac{\tan\theta_1 + K\tan\theta_2}{Z_2\tan\theta_1\tan\theta_2 - Z_1}, \tag{3.8}$$

where $\theta_1 = \beta_1 l_1$ and $\theta_2 = \beta_2 l_2$ are the electrical lengths of the low- and high-impedance sections (β denoting the phase constant), respectively, and

[15]The discontinuities (the step in width and the open-ended) are neglected in this expression, although can be included through their equivalent circuit models [18].

$$K = \frac{Z_1}{Z_2}, \tag{3.9}$$

is an important parameter named impedance ratio (or contrast) of the SISS [42]. At resonance, the denominator in (3.8) vanishes, and the following condition results

$$K = \tan\theta_1(\omega_0)\tan\theta_2(\omega_0), \tag{3.10}$$

where $\omega_0 = \sqrt{LC}$ is the angular resonance frequency. It is obvious from (3.10) that to minimize the total electrical length ($\theta_T = \theta_1 + \theta_2$) of the resonator, the impedance ratio must be $K < 1$. Furthermore, the higher the impedance contrast ($K << 1$), the smaller the electrical length of the SISS [42]. The reduction of θ_T is important for miniaturization and to be able to describe the SISS by means of a lumped element model. Therefore, if the impedance contrast is high enough, the electrical lengths of the resonator sections may attain small sizes at resonance. Accordingly, the SISS may be referred to as a planar semi-lumped or even lumped resonator [18].

Under the assumption that the two transmission-line sections of the SISS are electrically small, we can linearize the tangents in (3.8), and the admittance of the SISS reduces to

$$Y_{\mathrm{SISS}} = -j\frac{\theta_1 + K\theta_2}{Z_2\theta_1\theta_2 - Z_1}. \tag{3.11}$$

This admittance is identical to that of a series LC resonant tank, given by

$$Y_{LC} = -j\frac{\omega C}{LC\omega^2 - 1}, \tag{3.12}$$

provided the following mapping is satisfied

$$C = \frac{l_1}{v_{p1}Z_1} + \frac{l_2}{v_{p2}Z_2} = C_1 + C_2, \tag{3.13a}$$

$$L = \frac{Z_2 l_2}{v_{p2}}\frac{\frac{l_1}{v_{p1}Z_1}}{\frac{l_1}{v_{p1}Z_1} + \frac{l_2}{v_{p2}Z_2}} = L_2\frac{C_1}{C_1 + C_2}, \tag{3.13b}$$

where v_{p1} and v_{p2} ($\beta = \omega/v_p$) are the phase velocities of the low- and high-impedance transmission-line sections, respectively, C_1 and C_2 are the line capacitances, namely,

$$C_1 = \frac{l_1}{v_{p1}Z_1} = C_{\mathrm{pul1}}l_1, \tag{3.14a}$$

$$C_2 = \frac{l_2}{v_{p2}Z_2} = C_{\mathrm{pul2}}l_2, \tag{3.14b}$$

$C_{\text{pul}i}$ ($i = 1, 2$) being the per-unit-length capacitances of the lines, and L_2 is the inductance of the high-impedance section, i.e.

$$L_2 = L_{\text{pul2}} l_2. \tag{3.15}$$

The mapping in (3.13) indicates that, provided that the two sections of the SISS are electrically small, the SISS distributed behavior may be approximated to that of a shunt-connected series LC resonator to a high degree. As is well-known, in a first-order approximation, the inductance/capacitance of the SISS is mainly determined by the inductance/capacitance of the high/low impedance section [15, 18]. Indeed, expressions (3.13) are accurate and simple closed-form expressions of the SISS inductance and capacitance. Hence, according to (3.13), note that although C is dominated by C_1, the contribution of C_2 may be non-negligible if either K or l_2 are not very small. It is also interesting to realize that L is not purely equal to L_2 but somewhat smaller. Obviously, in a second level approximation, $C \approx C_1$ and $L \approx L_2$, but this approximation sacrifices accuracy and it is not actually necessary for our purposes.

To determine the implementable LC values, θ_1 and θ_2 are expressed as a function of L and C. The following results are obtained

$$\theta_1 = \frac{C\omega Z_1}{2}\left(1 + \sqrt{1 - \frac{4L}{CZ_2^2}}\right), \tag{3.16a}$$

$$\theta_2 = \frac{C\omega Z_2}{2}\left(1 - \sqrt{1 - \frac{4L}{CZ_2^2}}\right). \tag{3.16b}$$

Since the region of interest is beyond the fundamental resonance, we can express the maximum frequency as $\omega_m = n\omega_0$ in (3.16), with $n > 1$. This leads to

$$\theta_1(\omega_m) = \frac{nZ_1}{2Z_{eq}}\left(1 + \sqrt{1 - 4\left(\frac{Z_{eq}}{Z_2}\right)^2}\right), \tag{3.17a}$$

$$\theta_2(\omega_m) = \frac{nZ_2}{2Z_{eq}}\left(1 - \sqrt{1 - 4\left(\frac{Z_{eq}}{Z_2}\right)^2}\right), \tag{3.17b}$$

where

$$Z_{eq} = \sqrt{\frac{L}{C}}. \tag{3.18}$$

Note that expressions (3.17) depend on L and C through Z_{eq}, i.e. they do not depend on the specific values of L and C, but on its ratio.

To guarantee that the SISS can be described by a simple lumped series LC resonator, we impose the electrical lengths of the SISS sections to be smaller than a certain maximum predetermined value, θ_m. As usual, we set $\theta_m = \pi/4$.[16] By forcing (3.17) to be smaller than $\pi/4$, i.e.

$$\theta_1(\omega_m) = \frac{nZ_1}{2Z_{eq}}\left(1 + \sqrt{1 - 4\left(\frac{Z_{eq}}{Z_2}\right)^2}\right) < \theta_m, \tag{3.19a}$$

$$\theta_2(\omega_m) = \frac{nZ_2}{2Z_{eq}}\left(1 - \sqrt{1 - 4\left(\frac{Z_{eq}}{Z_2}\right)^2}\right) < \theta_m, \tag{3.19b}$$

the implementable values of Z_{eq} can be obtained.

The allowable LC values are dictated by the limits on the minimum and maximum implementable characteristic impedances. In order to enhance the range of implementable LC values, it is necessary to deal with the extreme (maximum and minimum) characteristic impedances that can be achieved with the considered substrate and fabrication technology. The maximum achievable impedance is that corresponding to the narrowest possible line, which is subjected to the fabrication process. Typically, for printed circuit boards (PCBs), the strip width cannot be narrower than 150 μm. On the other hand, the lowest impedance limit is not given by any technological constraint, but the width of the low-impedance section is set to the maximum width that avoids the appearance of transverse-resonant modes at $\omega_m = n\omega_0$, namely [46]

$$w_{1\,\max} = \frac{c\pi}{n\omega_0\sqrt{\varepsilon_r}} - 0.4h, \tag{3.20}$$

where c is the speed of light in vacuum, h is the substrate thickness and ϵ_r is the substrate relative permittivity. From (3.20) and ω_0 (given by the LC values), $Z_1(\omega_0)$ can be determined, which is certainly frequency-dependent.

Fig. 3.23 represents the allowable values of Z_{eq} under different scenarios. As f_0 increases, the allowable interval of Z_{eq} decreases, the upper limit being constant since it is determined by the frequency-independent θ_2. Next, the information in Fig. 3.23 is translated to the LC plane, that is, each f_0 corresponds to a certain curve in the LC plane. From the range of values of Z_{eq} corresponding to a particular f_0, the allowable values of L and C in each curve are obtained, as plotted in Fig. 3.24. The representation is limited to capacitances of 12 pF and inductances of 12 nH since higher values give rise to extremely large SISS dimensions for being practical. However, from a theoretical point of view, very small values of f_0 with implementable LC values can be obtained. By contrast, as f_0 increases, the allowable interval of LC values decreases dramatically, and finally collapses.[17] The width of the capacitive section, which must decrease to avoid the transverse-resonance mode, is the respon-

[16] A transmission line is typically considered to be electrically short if $\theta < \pi/4$, but we can specify whatever arbitrary maximum permissible electrical length.

[17] There is a fundamental limit on the maximum resonance frequency, but structure scaling down is possible if necessary.

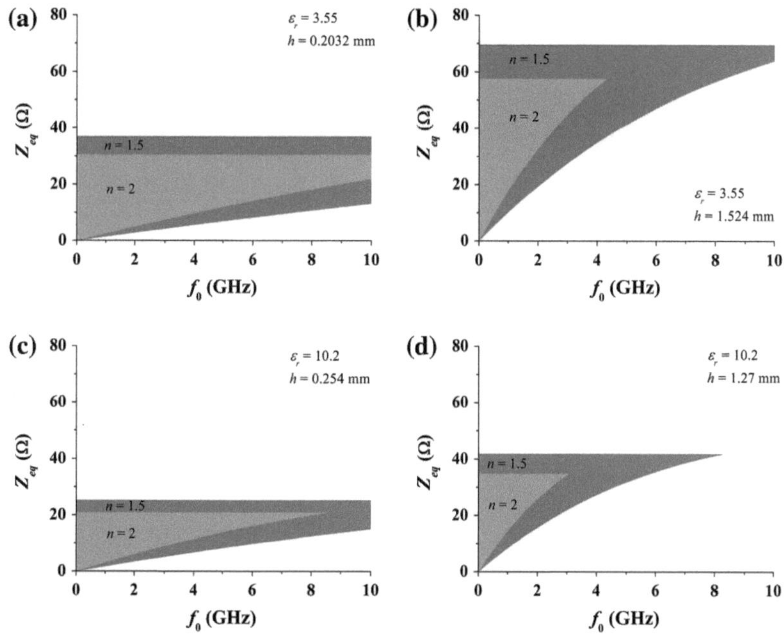

Fig. 3.23 Implementable values of Z_{eq} as a function of the resonance frequency for four different commercially available microwave substrates (with thickness h and relative permittivity ϵ_r) for $n = 1.5$ and $n = 2$. **a** $h = 0.203$ mm and $\epsilon_r = 3.55$. **b** $h = 1.524$ mm and $\epsilon_r = 3.55$. **c** $h = 0.254$ mm and $\epsilon_r = 10.2$. **d** $h = 1.27$ mm and $\epsilon_r = 10.2$. The minimum strip width considered is $w_2 = 150$ μm. Reprinted with permission from [45]

sible for this frequency limitation (at high frequencies the capacitance is limited to small values to avoid electrically large capacitive lines).

Finally, to validate the previous analysis, assume the range in Fig. 3.24d for $n = 2$. Next, consider a pair of LC values, one of them within the allowable region in the LC plane (*case A*), and another one that does not belong to the allowable region (*case B*). The *case A* is $L = 3.7$ nH and $C = 6.9$ pF where $f_0 = 0.996$ GHz; the *case B* is $L = 7.2$ nH and $C = 3.5$ pF where the resonance frequency is roughly the same, $f_0 = 1.003$ GHz. The narrowest line width is set to $w_2 = 150$ μm ($Z_2 = 100.9$ Ω). From (3.20), $w_1 = 23$ mm ($Z_1 = 5.8$ Ω). For *case A*, the lengths have been determined from (3.13)[18]; $l_1 = 2.8$ mm and $l_2 = 4$ mm ($\theta_1 = 20.7°$ and $\theta_2 = 23.8°$ at $2f_0$). For *case B*, the reported design procedure can be no longer used, and the line lengths are determined by curve fitting the reactance slope: $l_1 = 0.65$ mm and $l_2 = 8.8$ mm ($\theta_1 = 4.8°$ and $\theta_2 = 52.4°$ at $2f_0$). A photograph of a 50-Ω line loaded with the synthesized SISSs is shown in Fig. 3.25.

The simulated and measured frequency responses of the considered SISS-loaded lines are plotted in Fig. 3.26. Furthermore, for comparison purposes, the response

[18]Some optimization is actually required due to the effects of discontinuities, not accounted for by the model.

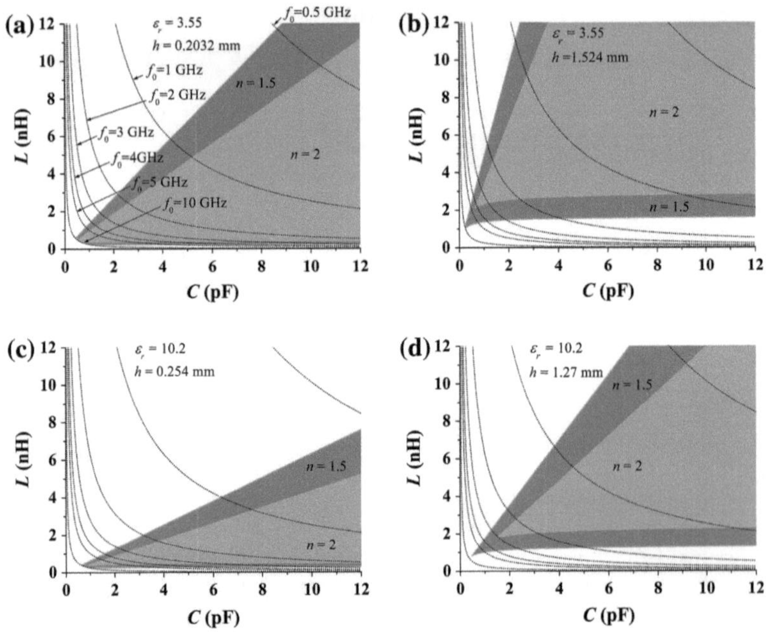

Fig. 3.24 Implementable values of L and C as a function of the resonance frequency for four different substrates for $n = 1.5$ and $n = 2$. **a** $h = 0.203$ mm and $\epsilon_r = 3.55$. **b** $h = 1.524$ mm and $\epsilon_r = 3.55$. **c** $h = 0.254$ mm and $\epsilon_r = 10.2$. **d** $h = 1.27$ mm and $\epsilon_r = 10.2$. Reprinted with permission from [45]

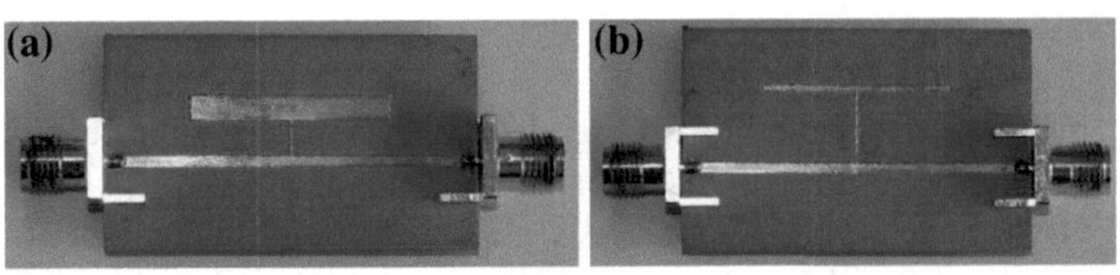

Fig. 3.25 SISS-loaded microstrip line corresponding to **a** *case A* and **b** *case B*. The substrate is *Rogers RO3010* with $h = 1.27$ mm and $\epsilon_r = 10.2$. Dimensions of the host microstrip line are: $W = 1.18$ mm (50-Ω line) and $l = 43$ mm. SISS widths are: $w_1 = 23$ mm and $w_2 = 150$ μm. SISS lengths are: for *case A*, $l_1 = 2.8$ mm and $l_2 = 4$ mm; for *case B*, $l_1 = 0.65$ mm and $l_2 = 8.8$ mm

of the circuit simulation of the same line loaded with the corresponding lumped shunt-connected LC resonator is also included. For *case A*, the agreement is excellent up to $2f_0$, confirming the expectations.[19] Hence, a highly accurate modeling is obtained when the inductance and capacitance belong to the inferred physically implementable values. Moreover, the transverse resonance mode is clearly visible

[19]The agreement is excellent between simulations, since measurements cannot be used as a proper reference on account of the usual fabrication-related deviations from ideal nominal physical and electrical parameters.

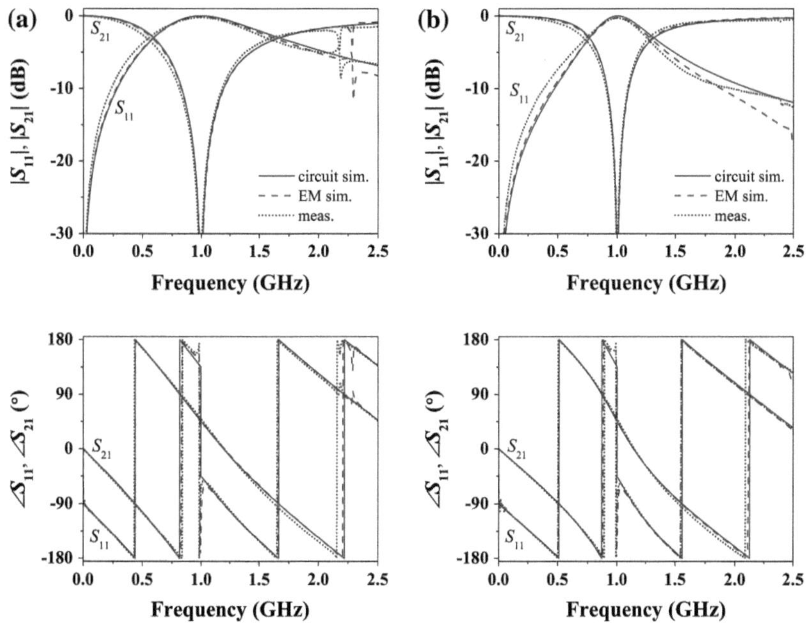

Fig. 3.26 Transmission and reflection coefficients, S_{21} and S_{11}, of the SISS-loaded microstrip lines in Fig. 3.25. **a** *Case A*. **b** *Case B*

beyond the frequency range of interest. Contrary to *case A*, in *case B* the SISS does not properly describe the behavior of the shunt connected LC series resonator up to $2f_0$. At this frequency, the discrepancy between the circuit and electromagnetic simulations in $|S_{11}|$ is clearly visible, and the phases of S_{11} and S_{21} differ about 2 or 3° (whereas for *case A*, the inaccuracy in phase is smaller than 1°). The reason is that the length of the high impedance line for *case B* cannot be considered electrically small. Further discrepancies are expected for cases where the capacitive and the inductive sections are both electrically large.

References

1. R. Marqués, F. Martín, M. Sorolla, *Metamaterials with Negative Parameters: Theory, Design and Microwave Applications*. (Wiley, New York, 2008)
2. R. Marqués, F. Medina, R. Rafii-El-Idrissi, Role of bianisotropy in negative permeability and left-handed metamaterials. Phys. Rev. B **65**(144440) (2002)
3. L. Solymar, E. Shamonina, *Waves in Metamaterials*. (Oxford University Press, Oxford, 2009)
4. P. Gay-Balmaz, O.J.F. Martin, Electromagnetic resonances in individual and coupled split-ring resonators. J. Appl. Phys. **92**(5), 2929–2936 (2002)
5. N. Katsarakis, T. Koschny, M. Kafesaki, E. Economou, C. Soukoulis, Electric coupling to the magnetic resonance of split ring resonators. Appl. Phys. Lett. **84**(15), 2943–2945 (2004)

6. J.D. Baena, J. Bonache, F. Martín, R.M. Sillero, F. Falcone, T. Lopetegi, M.A.G. Laso, J. García-García, I. Gil, M.F. Portillo, M. Sorolla, Equivalent-circuit models for split-ring resonators and complementary split-ring resonators coupled to planar transmission lines. IEEE Trans. Microw. Theory Tech. **53**(4), 1451–1461 (2005)
7. J. Naqui, M. Durán-Sindreu, F. Martín, Modeling split-ring resonator (SRR) and complementary split-ring resonator (CSRR) loaded transmission lines exhibiting cross-polarization effects. IEEE Antennas Wireless Propag. Lett. **12**, 178–181 (2013)
8. J. Naqui, M. Durán-Sindreu, F. Martín, Selective mode suppression in coplanar waveguides using metamaterial resonators. Appl. Phys. A **109**(4), 1053–1058 (2012)
9. J. Naqui, M. Durán-Sindreu, F. Martín, Selective mode suppression in microstrip differential lines by means of electric-LC (ELC) and magnetic-LC (MLC) resonators. Appl. Phys. A **115**, 637–643 (2014)
10. F. Martín, J. Bonache, F. Falcone, M. Sorolla, R. Marqués, Split ring resonator-based left-handed coplanar waveguide. Appl. Phys. Lett. **83**(22), 4652–4654 (2003)
11. F. Martín, F. Falcone, J. Bonache, R. Marqués, M. Sorolla, Miniaturized coplanar waveguide stop band filters based on multiple tuned split ring resonators. IEEE Microw. Wireless Compon. Lett. **13**(12), 511–513 (2003)
12. F. Aznar, J. Bonache, F. Martín, Improved circuit model for left-handed lines loaded with split ring resonators. Appl. Phys. Lett. **92**(043512) (2008)
13. F. Aznar, M. Gil, J. Bonache, L. Jelinek, J. D. Baena, R. Marqués, F. Martín, Characterization of miniaturized metamaterial resonators coupled to planar transmission lines through parameter extraction. J. Appl. Phys. **104**(114501) (2008)
14. J. Bonache, M. Gil, I. Gil, J. García-García, F. Martín, On the electrical characteristics of complementary metamaterial resonators. IEEE Microw. Wireless Compon. Lett. **16**(10), 543–545 (2006)
15. D.M. Pozar, *Microwave Engineering*, 3rd edn. (Wiley, New York, 2005)
16. R. Marqués, J. Baena, J. Martel, F. Medina, F. Falcone, M. Sorolla, F. Martín, Novel small resonant electromagnetic particles for metamaterial and filter design, in *International Conference on Electromagnetic in Advanced Applications (ICEAA '03)*, (Torino, Italy, 2003), pp. 439–442
17. W. Hayt, J. Kemmerly, S. Durbin, *Engineering Circuit Analysis*. (McGraw-Hill, New York, 2011)
18. J.-S. Hong, M.J. Lancaster, *Microstrip Filters for RF/Microwave Applications*. (Wiley, New York, 2001)
19. F. Hesmer, E. Tatartschuk, O. Zhuromskyy, A.A. Radkovskaya, M. Shamonin, T. Hao, C.J. Stevens, G. Faulkner, D.J. Edwards, E. Shamonina, Coupling mechanisms for split ring resonators: theory and experiment. Phys. Status Solidi B **244**(4), 1170–1175 (2007)
20. E. Tatartschuk, N. Gneiding, F. Hesmer, A. Radkovskaya, E. Shamonina, Mapping inter-element coupling in metamaterials: Scaling down to infrared. J. Appl. Phys. **111**(094904) (2012)
21. B.R. Bojanic, V. Milosevic, On the orientation of split-ring resonators excited by guided waves, in *Metamaterials '2012: The 6th International Congress on Advanced Electromagnetic Materials in Microwaves and Optics* (St. Petersburg, Russia, 2012)
22. I. Awai, Y. Zhang, Separation of coupling coefficient between resonators into magnetic and electric components toward its application to BPF development, in *China-Japan Joint Microwave Conference* pp. 61–65 (2008)
23. R.R. Syms, E. Shamonina, V. Kalinin, L. Solymar, A theory of metamaterials based on periodically loaded transmission lines: Interaction between magnetoinductive and electromagnetic waves. J. Appl. Phys. **97**(064909) (2005)
24. R.R.A. Syms, L. Solymar, Effective permeability of a metamaterial: against conventional wisdom. Appl. Phys. Lett. **100**(124103) (2012)
25. J. Neto, J. Barroso, P. Castro, Dispersion relation of split-ring resonators magnetically coupled to a TEM wave, in *SBMO/IEEE MTT-S International Microwave and Optoelectronics Conference (IMOC)*, (Rio de Janeiro , Brazil, 2013) pp. 1–5

26. M. Mrozowski, J. Mazur, Matrix theory approach to complex waves [in shielded lossless guides]. IEEE Trans. Microw. Theory Tech. **40**(4), 781–785 (1992)
27. T. Tamir, A. Oliner, Guided complex waves. Proc. Inst. Elect. Eng. **110**(2), 310–334 (1963)
28. W.-X. Huang, T. Itoh, Complex modes in lossless shielded microstrip lines. IEEE Trans. Microw. Theory Tech. **36**(1), 163–165 (1988)
29. M.J. Freire, F. Mesa, M. Horno, Excitation of complex and backward mode on shielded lossless printed lines. IEEE Trans. Microw. Theory Tech. **47**(7), 1098–1105 (1999)
30. F. Elek, G. Eleftheriades, Dispersion analysis of the shielded sievenpiper structure using multiconductor transmission-line theory. IEEE Microw. Wireless Compon. Lett. **14**(9), 434–436 (2004)
31. R. Islam, G. Eleftheriades, On the independence of the excitation of complex modes in isotropic structures. IEEE Trans. Antennas Propag. **58**(5), 1567–1578 (2010)
32. J. Naqui, M. Durán-Sindreu, F. Martín, A. Fernández-Prieto, F. Mesa, F. Medina, Complex modes in periodic transmission lines based on split rings, in *International Conference on Electromagnetics in Advanced Applications (ICEAA '13)*, (Torino, Italy, 2013)
33. J. Naqui, A. Fernández-Prieto, F. Mesa, F. Medina, F. Martín, Effects of inter-resonator coupling in split ring resonator loaded metamaterial transmission lines. J. Appl. Phys. **115**(194903) (2014)
34. J. Naqui, M. Durán-Sindreu, A. Fernández-Prieto, F. Mesa, F. Medina, F. Martín, Multimode propagation and complex waves in CSRR-based transmission-line metamaterials. IEEE Antennas Wireless Propag. Lett. **11**, 1024–1027 (2012)
35. R.K. Mongia, J. Hong, P. Bhartia, I.J. Bahl, *RF and Microwave Coupled-Line Circuits*. (Artech House, Boston, 1999)
36. J. Naqui, F. Martín, Transmission lines loaded with bisymmetric resonators and their application to angular displacement and velocity sensors. IEEE Trans. Microw. Theory Tech. **61**(12), 4700–4713 (2013)
37. R. Islam, M. Zedler, G. Eleftheriades, Modal analysis and wave propagation in finite 2D transmission-line metamaterials. IEEE Trans. Antennas Propag. **59**(5), 1562–1570 (2011)
38. C.-H. Tsai, T.-L. Wu, A broadband and miniaturized common-mode filter for gigahertz differential signals based on negative-permittivity metamaterials. IEEE Trans. Microw. Theory Tech. **58**(1), 195–202 (2010)
39. I. Gil, J. Bonache, M. Gil, J. García-García, F. Martín, R. Marqués, Accurate circuit analysis of resonant-type left handed transmission lines with inter-resonator coupling. J. Appl. Phys. (074908) (2006)
40. J. Selga, Synthesis of microwave circuits based on metamaterials using aggressive space mapping algorithms. Ph.D. dissertation, Universitat Autònoma de Barcelona (2013)
41. J. Selga, J. Naqui, M. Durán-Sindreu, F. Martín, A. Rodríguez, V. Boria, Application of aggressive space mapping (ASM) to the efficient synthesis of stepped impedance resonators (SIRs), in *European Microwave Conference (EuMC)*, (Nuremberg, Germany, 2013) pp. 636–639
42. M. Makimoto, S. Yamashita, Compact bandpass filters using stepped impedance resonators. Proc. IEEE **67**(1), 16–19 (1979)
43. J.-K. Lee, D.-H. Lee, Y.-S. Kim, A compact low-pass filter with double-step impedance shunt stub and defected ground structure for wideband rejection. Microw. Opt. Technol. Lett. **52**(1), 132–134 (2010)
44. J. Park, J.-P. Kim, S. Nam, Design of a novel harmonic-suppressed microstrip low-pass filter. IEEE Microw. Wireless Compon. Lett. **17**(6), 424–426 (2007)
45. J. Naqui, M. Durán-Sindreu, J. Bonache, F. Martín, Implementation of shunt-connected series resonators through stepped-impedance shunt stubs: analysis and limitations. IET Microw. Antennas Propag. **5**(11), 1336–1342 (2011)
46. T.C. Edwards, M.B. Steer, *Foundations of Interconnect and Microstrip Design*. (Wiley, New York, 2000)

Chapter 4
On the Symmetry Properties of Resonator-Loaded Transmission Lines

This chapter is focused on the symmetry properties of transmission lines loaded with electrically small resonators. The transmission characteristics of these structures are derived in terms of symmetry. It is shown that, by breaking symmetry, controllability in the signal transmission is obtained. Two different approaches are considered which achieve different transmission controllability, depending on whether the line is loaded with single or pairs of resonators (per-unit-cell). In the former, symmetric transmission lines are loaded with single symmetric resonators. Since no other reactive elements (neither inductances nor capacitances) are loaded to the line, one should expect to obtain stopband responses (Sect. 2.2.4). However, it may be obtained not only stopband responses, but also all-pass responses. As will be shown, this can be achieved by inhibiting resonance. In the latter, symmetric transmission lines are loaded with pairs of identical resonators. In these structures resonance always occurs, but controllability in the frequency of the resonances is performed.

The organization of this chapter is as follows. Section 4.1 describes some typical transmission lines that are used throughout this thesis, with special emphasis on their symmetry properties of the fundamental propagation modes. Specifically, microstrip lines, pairs of coupled microstrip lines, and coplanar waveguides are considered. Electrically small resonators are beyond the scope of this chapter because they were already introduced in Sect. 2.3. Sections 4.2 and 4.3 are devoted to the particular case of alignment and misalignment of symmetry planes in the former approach, respectively. Finally, Sect. 4.4 generalizes the behavior of the two considered approaches as a function of symmetry properties.

4.1 On the Symmetry Properties of Transmission Lines

Figure 4.1 shows the three symmetric transmission lines that are utilized throughout this thesis: microstrip lines, pairs of microstrip lines whose coupling depends on their separation, and coplanar waveguides (CPWs). Microstrip lines are two-conductor

J. Naqui, *Symmetry Properties in Transmission Lines Loaded with Electrically Small Resonators*, Springer Theses,
DOI 10.1007/978-3-319-24566-9_4

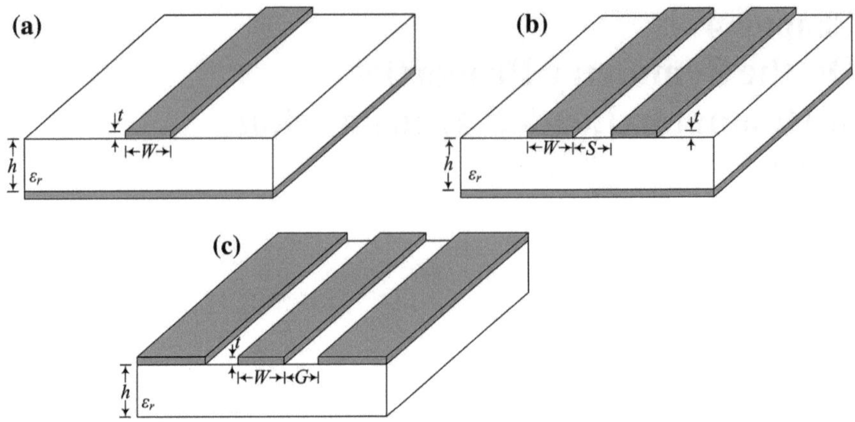

Fig. 4.1 Geometry of common transmission lines. **a** Microstrip line. **b** Pair of parallel symmetric microstrip lines. **c** Coplanar waveguide

transmission lines that support a single fundamental mode of propagation, whereas symmetric coupled microstrip lines and symmetric CPWs are three-conductor transmission lines allowing for the propagation of two fundamental modes, namely, the *even mode* and the *odd mode* [1]. Since part of the electromagnetic fields extends into the air beyond the substrate, the considered transmission lines are inhomogeneous propagative structures, and modes purely transverse electromagnetic (TEM) cannot be supported. Nonetheless, the propagation modes supported by these lines are of quasi-TEM nature [1], and are characterized by their characteristic impedance and propagation constant.[1] Therefore, the electric and magnetic fields are transverse to the direction of propagation to a considerable extent, the typical distribution of fields being represented in Fig. 4.2. As can be seen, in a microstrip line, there is a virtual magnetic wall at its symmetry plane. In coupled microstrip lines and CPWs, the symmetry plane behaves as a virtual electric or magnetic wall for the odd and even mode, respectively.[2]

In CPWs, the even and odd modes are alternatively characterized by the *CPW mode* (or *common mode*) and the *slot* (or *slotline* or *differential*) *mode*, respectively. Typically, the CPW mode is excited by feeding the central strip and grounding the lateral planes, whereas the slot mode is a parasitic mode[3] related to the signal across the slots and the central strip. For this reason, if the excitation mode is not

[1]A homogeneous dielectric material with an effective dielectric permittivity replaces the inhomogeneous medium.

[2]The electric/magnetic field components are normal to an/a electric/magnetic wall since the tangential components vanish [1].

[3]Pairs of microstrip lines, slotlines, or coplanar strips (CPS) are generally preferred to be fed by odd-mode sources [2, 3]. A slotline consists of a metallic layer where a slot is etched (see Fig. 4.12). When the two conductors are made finite, the structure is termed CPS, which can be viewed as the complementary of a CPW.

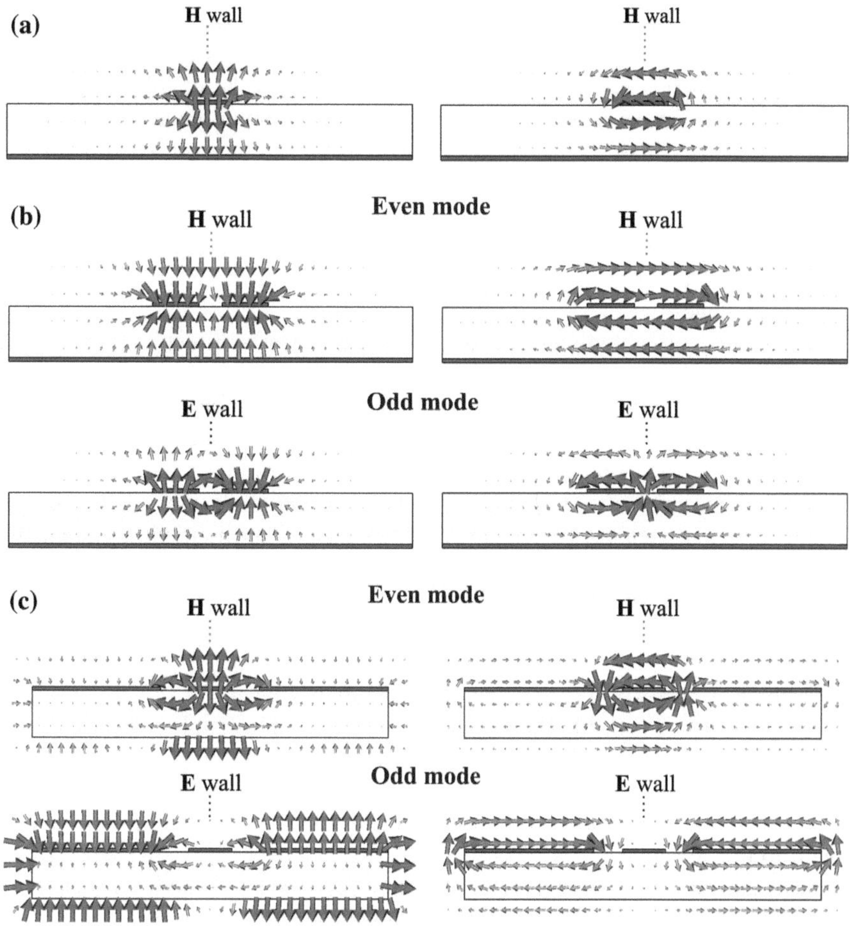

Fig. 4.2 Electric (*left*) and magnetic (*right*) field distribution in common transmission lines, obtained by the electromagnetic simulation software *Computer Simulation Technology (CST) Microwave Studio*. **a** Microstrip line. **b** Pair of coupled microstrip lines. **c** Coplanar waveguide, where a finite ground CPW (FGCPW) is considered. The boundary condition (electric or magnetic wall) at the symmetry plane is indicated

explicitly indicated, CPWs are always excited throughout this thesis with the common mode. Regarding coupled microstrip lines, they are useful for the design of differential microstrip lines. In this context, the odd and even mode are often alternatively defined by the *differential*- and *common-mode* (Appendix D). In contrast to CPWs, in differential circuits based on coupled microstrip lines, the undesired mode is the common mode.

In symmetric transmission lines, the propagation of the common- and differential-mode is independent, and power conversion from one mode to another does not arise. Therefore, a common-mode excitation cannot generate a differential-mode signal along the structure and vice versa. On the contrary, when a three-conductor transmission-line is asymmetric, the two fundamental modes are coupled to one another, and energy is continuously exchanged between the coupled modes (i.e. there is *mode conversion*) [4]. For example, from a common-mode source applied to an asymmetric CPW, the parasitic slot mode may be excited by absorbing power from the desired CPW mode. A common procedure to suppress the slot mode in CPWs consists of introducing air bridges to short circuit the ground planes at a particular transverse plane [5]. In printed circuit board (PCB) technology, via-holes in the substrate and strips printed on the back substrate side are usually used to short the ground planes.

When a network is symmetric, it is usually convenient to bisect it into two identical halves with respect to its symmetry plane, because the analysis of N-port networks is reduced to that of $N/2$-port networks using the *even/odd-mode analysis* (see Appendix C). Even and odd excitations apply in-phase and anti-phase signals, respectively. When an odd excitation is applied to a network, the symmetry plane is short-circuited, and the two halves become the same odd-mode network. In a similar fashion, under an even excitation, the symmetry plane is open-circuited and the two halves become the even-mode network. Therefore, in symmetric structures based on coupled microstrip lines and CPWs, one can restrict the analysis to half of the structure. Thereby two-port equivalent circuit models for the even and odd-mode can be derived.

Lastly, it is also worth discussing about the main advantages of the considered transmission lines [3, 5, 6]. The advantage of microstrip technology over CPWs is the presence of the ground plane on the back side of the substrate, which effectively isolates the structure preventing from possible interferences.[4] An advantage of single microstrip lines is that conversion between fundamental modes is not possible. With regard to the advantages of CPWs relative to microstrip lines, CPWs need only a single metallic layer and their characteristic impedance does not univocally determine the transverse dimensions (W and G) providing design flexibility.

4.2 On the Alignment of Symmetry Planes

Let us consider a symmetric transmission line loaded with a single symmetric resonator. Typically, a transmission line loaded with a resonator coupled to it exhibits a stopband (a transmission zero or notch, or an attenuation peak) at the fundamental resonance frequency (the magnitude of the transmission coefficient is $|S_{21}| \approx 0$). Suppression of the signal transmission occurs because the injected power is reflected back to the source. However, it is possible to prevent resonance and the appearance of the associated notch in the transmission coefficient.

[4]CPWs with backside metalization are known as conductor-backed CPWs (CB-CPWs), and we are forced to use it exceptionally in Sect. 6.7.1, but the inclusion of the metalization may give rise to parasitic modes [3].

The principle of operation behind the absence or presence of resonance[5] is based on the alignment of the symmetry planes of the line and the resonator as follows. *Assume that the symmetry plane of the resonator* (Sect. 2.3) *is perfectly aligned with the symmetry plane of the transmission line* (Sect. 4.1), *so that the structure is perfectly symmetric with regard to the longitudinally-oriented midplane.*

- *If the two symmetry planes are of the same sort* (both exhibiting electric or magnetic walls), *the resonator is excited as often. The structure is resonant and signal transmission is suppressed in the vicinity of the resonance frequency, resulting in a stopband response* ($|S_{21}| \approx 0$). *The resonator may be electrically, magnetically, or magnetoelectrically coupled to the line.*
- *If the two symmetry planes are of different sort* (one being a magnetic wall and the other one an electric wall), *the resonator cannot be excited. Because of perfect symmetry, the net electric and magnetic field components illuminating the resonator exactly cancel (the electric and magnetic fluxes over the resonator area are zero). Therefore, the resonance condition is inhibited because it can be established by neither electric nor magnetic coupling. Accordingly, the whole structure behaves as a transmission line exhibiting an all-pass response* ($|S_{21}| \approx 1$).[6]

It should be pointed out that the above-mentioned behavior holds regardless of the physical topologies of the transmission line and the resonator, and the line-to-resonator coupling mechanism (electric and/or magnetic). To clarify the previous assertions related to the resonance prevention, with no loss of generality, consider as an illustrative example the ELC-loaded CPW shown in Fig. 4.3. Such a transmission line exhibits a propagation mode with a magnetic wall at its symmetry plane. If one aligns with such a line a resonator that induces a magnetic wall at resonance, the resonator will be excited. On the contrary, in the case where there is an electric wall at the symmetry plane of the resonator, resonance is not allowed. Figure 4.4 shows the current distribution at the fundamental resonance of the ELC resonator for these two different alignments. It is apparent that, when the structure does not support resonant currents, most of the current flows on the CPW, whereas large currents are induced in the resonator when it is excited.

Now assume bimodal symmetric transmission lines such as coupled microstrip lines and CPWs. These lines support the propagation of the even and odd modes which exhibit walls of different nature (magnetic and electric walls, respectively). In the presence of a resonator loaded symmetrically to these lines, the electric and magnetic walls at the symmetry plane of the transmission lines are preserved for their corresponding modes. It follows that one can *selectively suppress either the even- or the odd-mode while preserving the other mode unaltered (or almost unaltered)*

[5]Although the fundamental resonance as well as higher-order resonances may be prevented, this thesis is focused on the fundamental resonance only.

[6]Obviously, the resonator may change the electrical parameters (i.e. the characteristic impedance and the phase constant) of the resulting propagative transmission line-type structure. In other words, the resonator may actually influence the transmission line parameters due to line-to-resonator coupling. Nevertheless, since resonance is effectively prevented, line-to-resonator coupling is said to be canceled throughout this thesis.

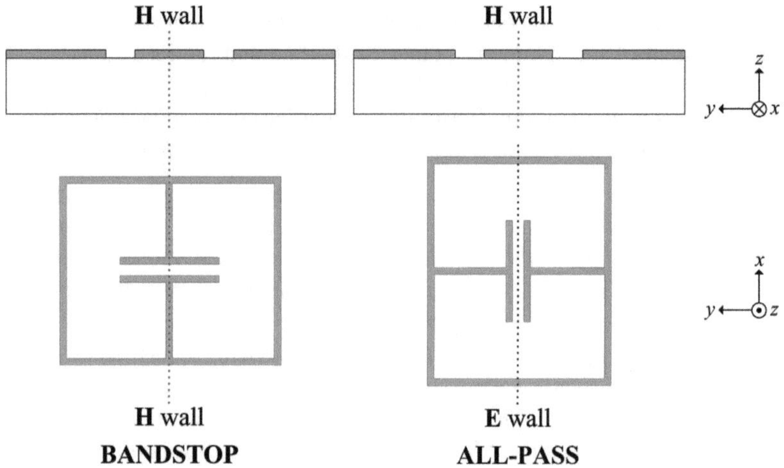

Fig. 4.3 Illustration of how resonance can be prevented. A CPW line loaded with an electric-*LC* (ELC) resonator etched on the bottom is considered

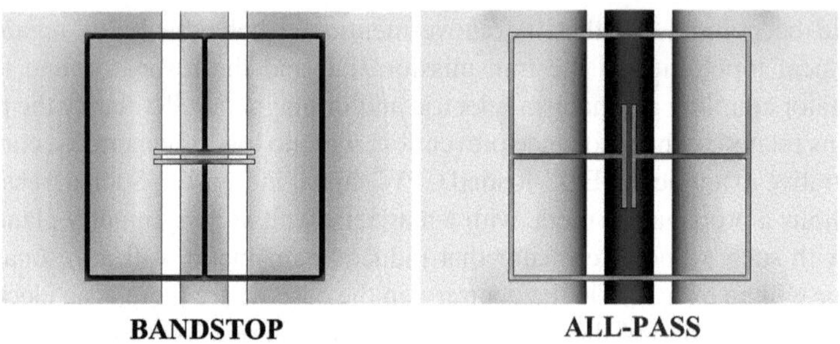

Fig. 4.4 Current distribution given by electromagnetic simulation depending on wether resonance is allowed or inhibited in an ELC-loaded CPW

(See footnote 6) *by aligning a resonator exhibiting a magnetic wall or an electric wall, respectively*. This filtering functionality, referred to as *selective mode suppression*, is illustrated in Fig. 4.5.

To the author's knowledge, the overlap of the symmetry plane of a resonator to that of a transmission line was applied for the first time in the context of differential lines with common mode suppression [7]. Indeed, in this thesis the alignment of symmetry planes is mainly exploited to design common-mode suppressed differential microstrip lines (Chap. 5). Nevertheless, the selective mode suppression is also verified in CPWs.

Next subsections are devoted to the considered implementations for achieving a selective mode suppression. Interestingly, since the structures are symmetric, the even/odd-mode analysis can be applied, and equivalent circuit models for even- and odd-mode excitations can be obtained (Appendix C). Alternatively, the

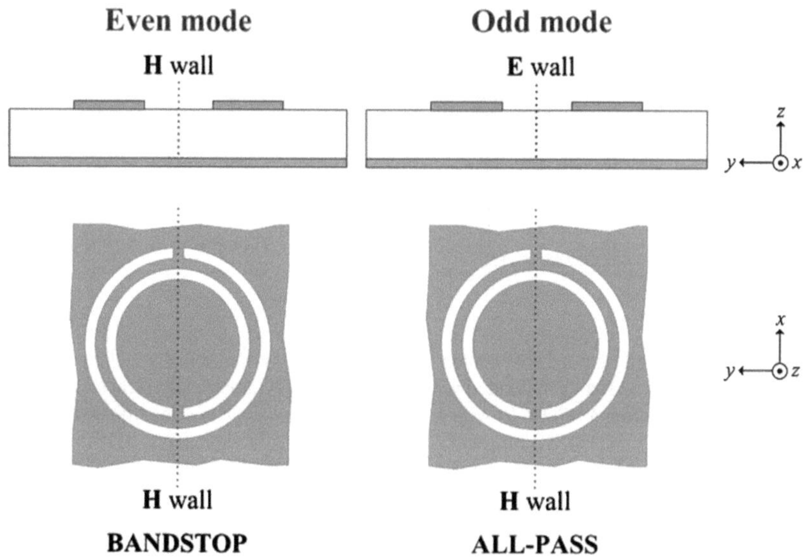

Fig. 4.5 Illustration of a selective mode suppression. A differential microstrip line loaded with a complementary split-ring resonator (CSRR) in the ground plane is considered

characterization can be performed by means of differential- and common-mode excitations (Appendix D).

4.2.1 SRR- and CSRR-Loaded Coplanar Waveguides

Coplanar waveguides loaded symmetrically with SRRs or CSRRs are analyzed in [8]. The SRRs are loaded on the back side of the substrate, whereas the CSRRs are etched in the CPW central strip (see Fig. 4.6). Symmetric loading is merely achieved by aligning the cuts of the rings with the line axis. Because the structures do not possess symmetry when looking into the ports, it is essential to identify the two ports (1 and 2). The CPW line will be excited by its two fundamental modes of propagation, and the port terminals (A, B, and C) are indicated for a better comprehension of the feeding schemes.

As shown in Fig. 4.7, CPWs symmetrically loaded with single (per-unit-cell) SRRs are transparent to signal propagation for the even (or common) mode, whereas they inhibit the odd (or differential) mode in the vicinity of resonance (equivalently to classical CPWs loaded with pairs of SRRs).[7] Conversely, CSRR-loaded CPWs are

[7]Typical CPWs loaded with pairs of SRRs inhibit the CPW (even) mode in the vicinity of the resonance frequency of the SRRs (Sect. 2.2.4). Unlike common SRR-loaded CPWs where the SRRs are etched below the slots, the considered SRR-loaded CPWs here are loaded with single SRRs below the CPW central strip.

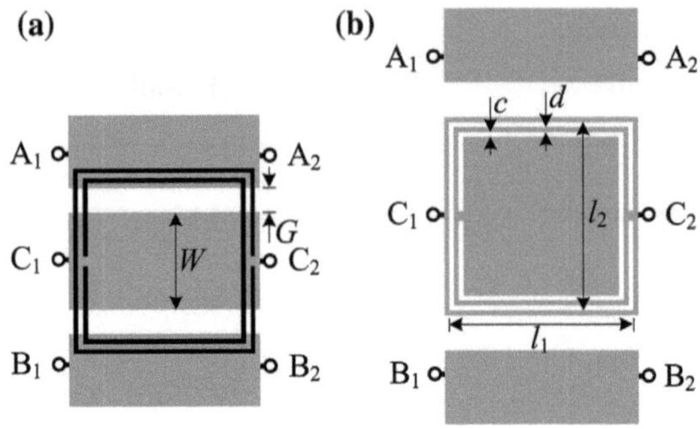

Fig. 4.6 Elemental cell of **a** a CPW loaded with single SRRs symmetrically etched on the back side of the substrate and **b** a CPW loaded with single CSRRs symmetrically etched in the central strip. The substrate is *Rogers RO3010* with $h = 1.27$ mm and $\epsilon_r = 10.2$. The dimensions are: for the SRR and CSRR, $c = d = 0.2$ mm and $l_1 = l_2 = 7.6$ mm; for the CPWs, **a** $W = 4$ mm and $G = 1$ mm, and **b** $W = 8$ mm and $G = 1.4$ mm (50-Ω line for the CPW mode)

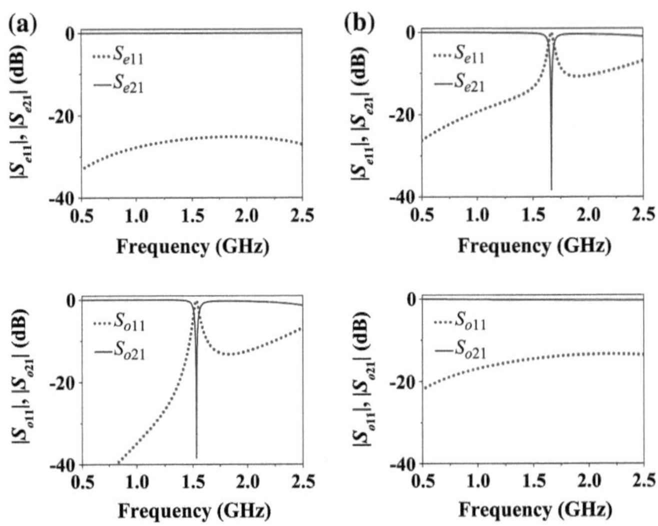

Fig. 4.7 Frequency response for the even (common) mode and odd (slot) mode of the CPWs loaded with **a** an SRR and **b** a CSRR in Fig. 4.6. The subscripts *e* and *o* stand for the even and odd modes, respectively. The common and differential modes are fed, respectively, by a 50-Ω port between terminals C and AB (A and B being shorted) and by a 100-Ω differential port between terminals A and B (C being left opened)

transparent for the slot mode, and opaque for the CPW mode in the vicinity of CSRR resonance (similar to CSRR-loaded microstrip lines). As was analyzed in detail in Sect. 3.1, due to the line-to-resonator orientation (necessary to guarantee symmetry longitudinally), mixed magnetoelectric coupling is present for the modes that are able to excite the fundamental resonance of the resonators.

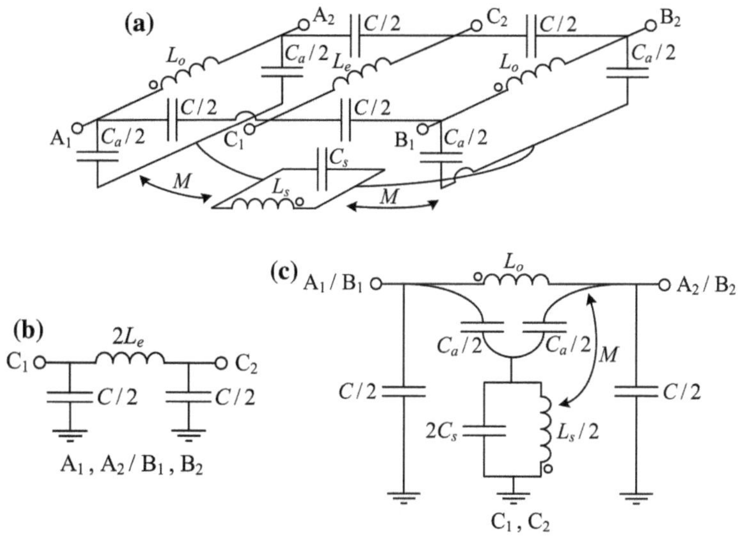

Fig. 4.8 **a** Six-terminal equivalent circuit model of a CPW loaded with symmetrically etched SRRs. Two-port **b** even- and **c** odd-mode equivalent circuit model. The capacitance C models the slot capacitance of the CPW, L_e and L_o are the inductance of the line for the even and odd modes, respectively, the SRR is modeled by L_s and C_s, M models the mutual inductance between the SRR and the CPW, and C_a accounts for the electric coupling between the line and the SRRs

Equivalent circuit models with magnetoelectric coupling of the considered SRR-loaded CPWs are provided in Fig. 4.8, including the equivalent circuits for even- and odd-mode excitations. For the common mode, the terminals A_1, A_2, B_1 and B_2 are grounded, a magnetic wall arises on the symmetry plane, the SRR is opened, and the equivalent circuit model is simply that of a conventional transmission line. For the differential mode, the feeding signal is applied between the terminals A_1 and B_1 (i.e. ports 1 and 2 are differential). Thus, the symmetry plane exhibits a virtual ground, and the equivalent circuit model that results after applying the electric wall concept is formally identical to that of a CPW loaded with a pair of SRRs.

By extracting the parameters of the odd-mode model from the procedure described in [10], the circuit simulation does not accurately fit to the full-wave simulation of the SRR-loaded CPW in Fig. 4.6 corresponding to the differential mode (as mentioned in Sect. 3.1, the reason is that such a procedure does not account for electric coupling). This suggests that electric coupling must be considered for an accurate description of the structure. Thus, the circuit values are extracted in a similar fashion as in Sect. 3.1 [11]. The extraction method is indicated in the caption of Fig. 4.9, and the comparison between the circuit and electromagnetic simulations is shown in this figure. As can be appreciated, good agreement is obtained by including electric coupling in the circuit simulation.

The reported lumped element equivalent circuit model of a CPW loaded with CSRRs in the central strip is shown in Fig. 4.10, including again the resulting

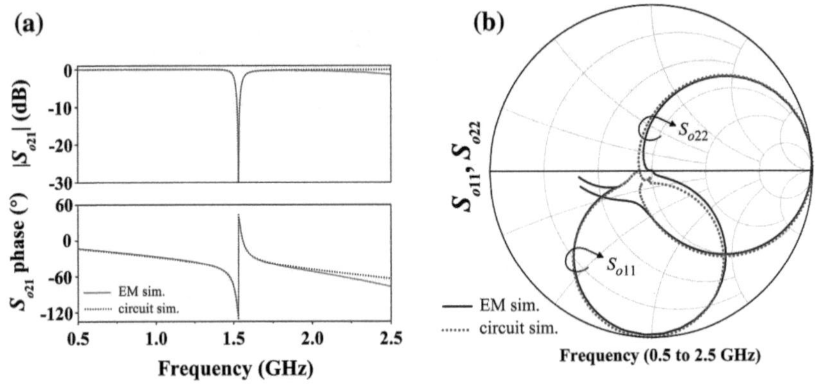

Fig. 4.9 **a** Transmission coefficient and **b** reflection coefficients of the SRR-loaded CPW in Fig. 4.6 for the odd mode. The elements of the circuit in Fig. 4.8c are extracted as follows. First, the elements of the transformed model of an SRR-loaded line in the absence of electric coupling are extracted from the procedure in [9] by aligning the slits of the SRR orthogonally to the line axis. Next, the coupling capacitance is included, and the circuit parameters in Fig. 4.8c are obtained by curve fitting (since neither L_s, C_s, nor M is calculated, there are an infinite number of mathematical solutions); $C = 1.38\,\text{pF}$, $L_o = 2.96\,\text{nH}$, $L_s = 3.66\,\text{nH}$, $C_s = 2.73\,\text{pF}$, $M = 0.66\,\text{nH}$, and $C_a = 0.4\,\text{pF}$

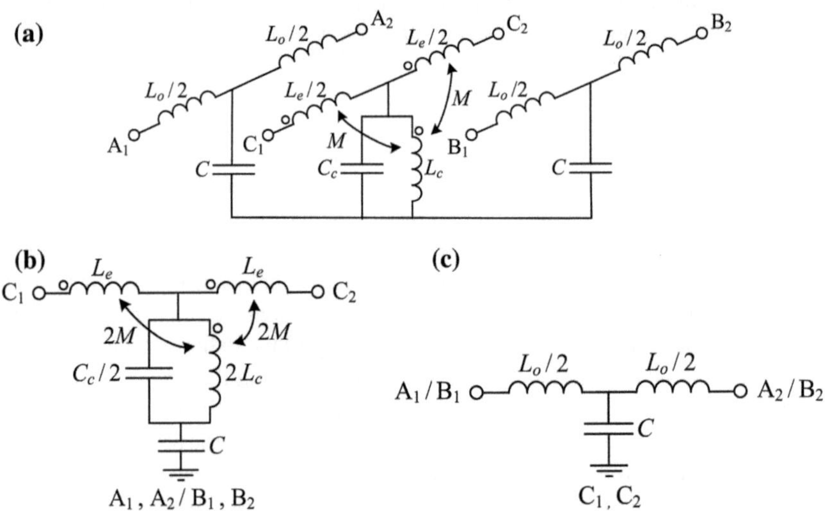

Fig. 4.10 **a** Six-terminal equivalent circuit model of a CPW loaded with symmetrically etched CSRRs. Two-port **b** even- and **c** odd-mode equivalent circuit model. The capacitance C models the slot capacitance of the CPW as well as the line-to-CSRR electric coupling, L_e and L_o are the inductance of the line for the even and odd modes, respectively, the CSRR is modeled by L_c and C_c, and M accounts for the magnetic coupling between the line and the CSRR

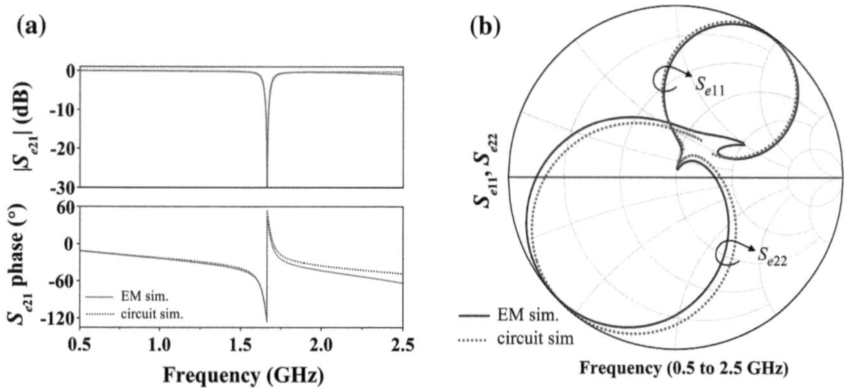

Fig. 4.11 **a** Transmission coefficient and **b** reflection coefficients of the CSRR-loaded CPW in Fig. 4.6 for the even mode. The elements of the circuit in Fig. 4.10b are: $C = 0.75$ pF, $L_e = 2.13$, $L_c = 0.49$ nH, $C_c = 17.3$ pF, and $M = 0.1$ nH. These elements are extracted by aligning the CSRR symmetry plane transversely to the line, and extracting the circuit elements (except M) by the procedure in [9]; then M is added and all the parameters are optimized to fit the electromagnetic simulation. The port impedance used in the circuit simulation is 50 Ω

equivalent circuits for even- and odd-mode excitations. For the odd mode, the CSRR is shorted to ground. Contrary to this and, similarly to the SRR-loaded CPW, the circuit model for the even mode corresponding to the CSRR-based CPW is formally equivalent to that of a CSRR-loaded microstrip line. Because of the CPW-to-CSRR mixed coupling, the circuit parameters for the even mode cannot be extracted using the procedure in [9]. Instead, the parameters for the even mode are extracted in the same manner as done in Sect. 3.1 [11]. Figure 4.11 plots the comparison between the electromagnetic and circuit simulations, from which it can be observed that both simulations are in good accordance.

The selective suppression of either the CPW (even) or the slot (odd) mode in CPWs using CSRRs and SRRs, respectively, is also supported by experimental results. In order to experimentally verify the selective mode suppression in CPWs, some symmetric structures are designed. One of them is shown in Fig. 4.12a and consists of a CPW loaded with an SRR on its back substrate side and with a CSRR in the central strip, while another structure is the same one without the SRR. The CPW ground planes are fed by means of a slotline to generate the slot mode, and it is clear from Fig. 4.12b that the presence of the SRR inhibits this mode at resonance (while this mode is not affected by the CSRR). On the other hand, two additional and identical structures without the slotline are also designed to excite the common mode: one of them with an SRR and a CSRR, as can be seen in Fig. 4.13a; in the other one, the CSRR is removed. As can be noticed in Fig. 4.13b, the situation is reversed, that is, the notch is caused by the CSRR, and the line is transparent at the resonance frequency of the SRR.

Finally, note that the response in Fig. 4.12b exhibits certain insertion losses due to the impedance mismatch produced at the connection between the coaxial cable of

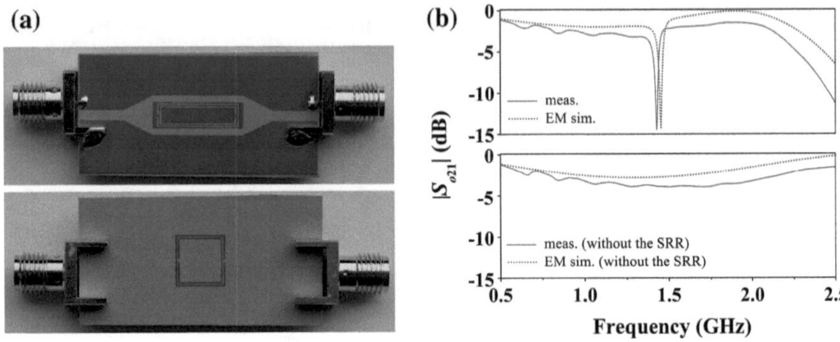

Fig. 4.12 **a** Top and bottom views of a CPW loaded with an SRR and a CSRR fed through a slotline to generate the slot mode, and **b** odd-mode transmission coefficient. The substrate is *Rogers RO3010* with $h = 1.27$ mm, $\epsilon_r = 11.2$, and loss tangent $\tan\delta = 0.0023$. The CPW and SRR dimensions are those for the SRR-loaded CPW in Fig. 4.6. The CSRR dimensions are: $c = d = 0.2$ mm, $l_1 = 12.6$ mm, and $l_2 = 3.6$ mm. The slotline width is 1.5 mm

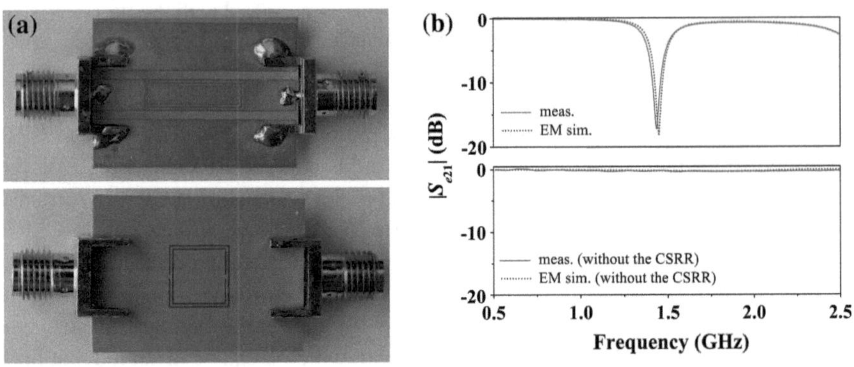

Fig. 4.13 **a** Top and bottom views of a CPW loaded with an SRR and a CSRR excited through the common mode, and **b** even-mode transmission coefficient. The structure is the same as that in Fig. 4.12 without the slotline

measurement and the slotline. The impedance of the former is 50 Ω, whereas that of the latter is found to be about 100 Ω. Nevertheless, this is irrelevant for the present purpose, since the main aim is to demonstrate the selective mode suppression of both even and odd modes in CPWs by using either SRRs or CSRRs. In this regard, it is important to mention that, for the sake of simplicity, the impedance of the ports in the even- and odd-mode circuits is simply set in all cases to 50 Ω (regardless of the open- or short-circuited boundary conditions applied to the symmetry plane).

CPW structures loaded with single SRRs or CSRRs may find applications in several fields. For which concern CSRR-loaded CPWs, the presence of these particles causes notches in the transmission coefficient for the fundamental mode, and this can be useful for the rejection of interfering signals in communication systems. As long as the CSRRs are etched in the central strip, this signal suppression can be achieved without the penalty of increasing device area and cost. Regarding SRR-loaded CPWs, a potential application might be the rejection of the parasitic slot mode that is generated in certain applications (namely, where a CPW-based structure is non-symmetric). In this thesis the potentiality of the slot-mode suppression by means of symmetric resonators has not been deeply investigated. However, a priori, it seems difficult to improve the performance of air bridges (or via-holes and strips), the standard solution to cancel the slot mode. Air bridges are very efficient structures for broadband slot-mode suppression, the CPW mode being practically unaltered. For this reason, vias and strips are used throughout this thesis to cancel the slot mode where asymmetric CPW-based structures are designed.

4.2.2 SRR- and CSRR-Loaded Differential Microstrip Lines

Either the odd- or the even-mode in differential microstrip lines can be selectively rejected using split-ring resonators (SRRs) or complementary split-ring resonators (CSRRs), respectively, in a similar way to CPWs. The resonators are loaded symmetrically by etching the SRRs between the pair of microstrip lines and the CSRRs in the ground plane, as illustrated in Fig. 4.14. The selective mode suppression using CSRRs and SRRs is demonstrated in Fig. 4.15 [12], where the reflection and transmission coefficients for the differential mode (S_{dd11} and S_{dd21}) and common mode (S_{cc11} and S_{cc21}) are plotted. As expected, an SRR is transparent for the common mode, whereas it suppresses the differential mode in the vicinity of resonance. Contrarily, a CSRR rejects the common mode, but it is transparent under the differential mode.

The possibility of rejecting the common mode while preserving the differential mode is of general interest in differential lines and circuits. In this regard, the selective

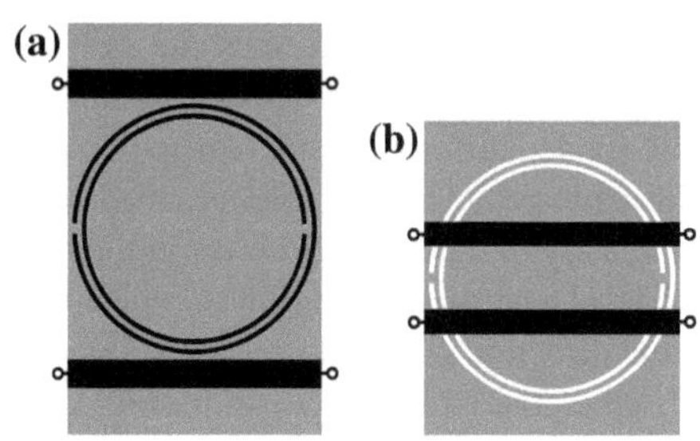

Fig. 4.14 **a** SRR- and **b** CSRR-loaded differential microstrip line for selective mode suppression. Resonator dimensions are (see Fig. 2.16): $c = d = 0.2$ mm and $r_{ext} = 5$ mm. Differential line dimensions are (exhibiting approximately an even- and odd-mode characteristic impedance of 50 Ω): for the SRR-loaded line, $W = 1.14$ mm and $S = 10.4$ mm; for the CSRR-loaded line, $W = 1$ mm and $S = 2.5$ mm. The substrate has a thickness of $h = 1.27$ mm and a relative permittivity of $\epsilon_r = 10.2$

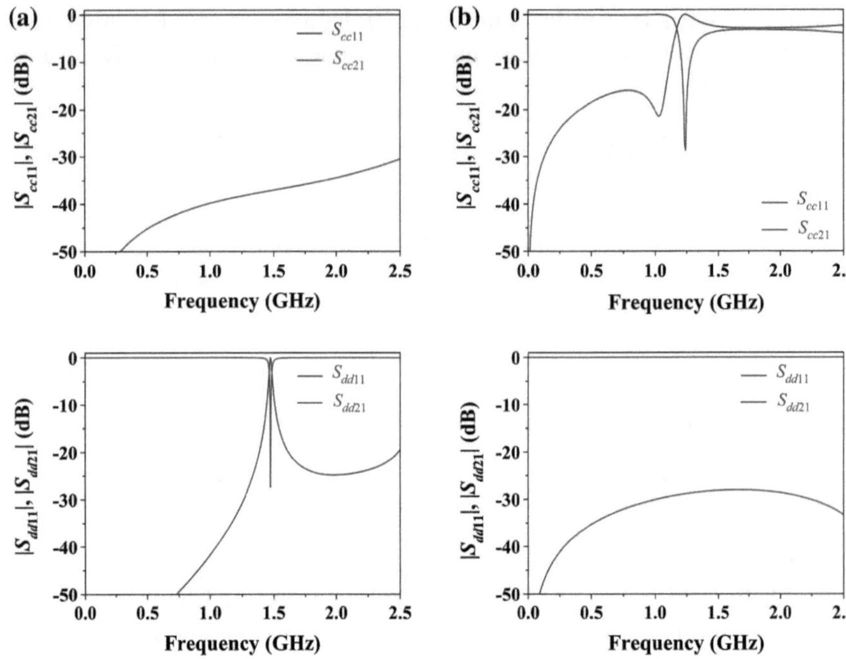

Fig. 4.15 Selective mode suppression in the **a** SRR- and **b** CSRR-loaded differential microstrip lines in Fig. 4.14

mode suppression in CSRR-loaded differential microstrip lines is analyzed for the first time in [13]. Subsequently, a comprehensive analysis was carried out in an extended work presented in [14]. The double-slit complementary split-ring resonator (DS-CSRR) is also considered therein as an alternative to the CSRR to enhance the common-mode rejection bandwidth (it follows that the DS-SRR can be used to suppress the differential mode). Chap. 5 presents the synthesis of some common-mode suppressed differential lines utilizing CSRRs, including experimental results. The behavior of a CSRR-loaded differential microstrip line can be explained in terms of electromagnetic fields from symmetry arguments as follows. Since for the common mode there is a strong density of electric field lines in the same direction below the microstrip lines, a net axial electric field excites the CSRRs producing a stopband. For the differential mode, the direction of the electric field lines is opposite in the strips of the differential line. Since the structure is symmetric, the opposite electric field vectors in both lines exactly cancel, there is no net axial electric field illuminating the CSRRs, and resonance is hence prevented. Analogously, a net magnetic field results oriented orthogonally to the CSRR symmetry plane for the common mode, but there is no net magnetic field in that direction for the differential mode. Therefore, neither of the two possible couplings (electric and magnetic) can excite the CSRRs in the differential mode, whereas there is line-to-resonator magnetoelectric coupling for the common mode.

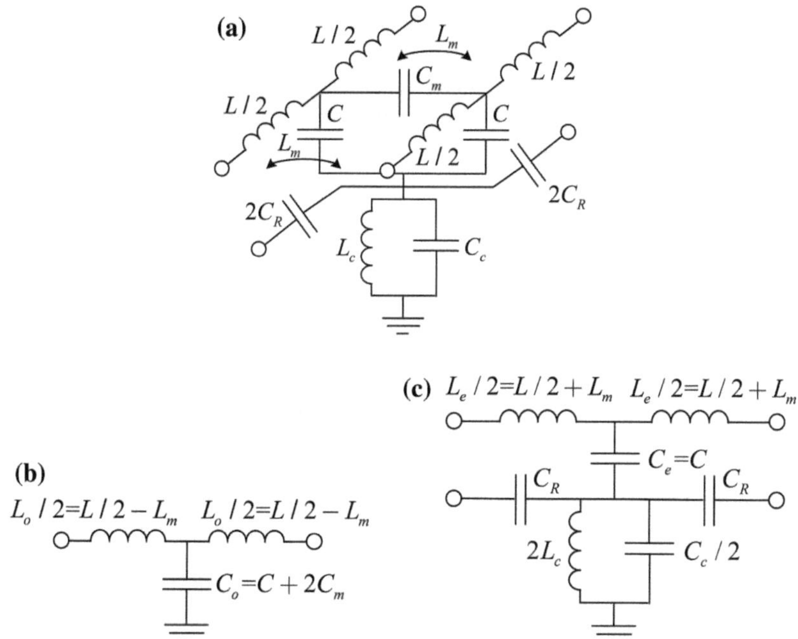

Fig. 4.16 Unit-cell equivalent circuit model of a microstrip differential line loaded with CSRRs symmetrically etched in the ground plane, including the inter-resonator capacitive coupling but neglecting mixed magnetoelectric coupling between the line and the CSRR. **a** Circuit of six single-ended ports of the whole structure. **b** Odd-mode circuit of two single-ended ports corresponding to half a structure. **c** Even-mode circuit of four single-ended ports corresponding to half a structure. The parameter L models the inductance of the lines, C accounts for the line capacitance and the electric coupling between the lines and the resonator, the CSRR is modeled by the parallel resonator, L_c and C_c, and C_m and L_m model the mutual capacitance and inductance between the coupled lines, respectively. Finally, C_R models the electric coupling between adjacent resonators

Note that mixed coupling is inevitable due to the CSRR orientation, dictated by symmetry requirements for the selective mode suppression. However, the line-to-resonator magnetic coupling is ignored for the sake of simplicity (using the DS-CSRR there is no magnetic coupling), and hence the dominant electric coupling is considered only.[8] An equivalent circuit model is presented in [14], shown in Fig. 4.16a, that includes the electric coupling between adjacent cascaded resonators, resulting in a six-port network using single-ended ports (such a coupling was neglected in the model presented previously in [13]). By feeding the circuit with an odd- or even-mode excitation, the simpler circuits in Figs. 4.16b and c result. For the odd mode, the resonator is short-circuited to ground, and the resulting model is the one of a conventional transmission line (all-pass response). For the even mode, the circuit is equivalent to that of a conventional CSRR-loaded microstrip line with inter-resonator coupling

[8]In the specific considered structures, mixed coupling is not significant for attenuation purposes since it basically changes the phase of the reflection coefficients (Sect. 3.1).

(stopband response). Hence, a four-port network (the same as in Sect. 3.2.1 but with modified CSRR parameters associated to half a resonator) is obtained, which is validated in Sect. 5.3.1. Evidently, all the results derived from the analysis of CSRR-loaded microstrip lines can be indeed extrapolated to CSRR-loaded microstrip differential lines under common-mode operation. In summary, the proposed model certainly explains that differential signals are insensitive to the presence of the CSRRs, while these resonators prevent the transmission of the common mode in the vicinity of CSRR resonance frequency.

4.2.3 ELC- and MLC-Loaded Differential Microstrip Lines

The selective mode suppression in differential lines has also been validated by means of ELC and MLC (electric and magnetic inductive-capacitive) resonators, the results being reported in [15, 16]. These resonators are bisymmetric and exhibit orthogonally oriented magnetic and electric walls at resonance. Therefore, the odd and even mode can be selectively suppressed by loading the lines with either ELC or MLC resonators (four different topological combinations are possible, two with even-mode suppression and the other two with odd-mode rejection).

Like in SRR- and CSRR-loaded differential lines, the ELC and MLC resonators are placed symmetrically between the microstrip lines and in the ground plane, respectively (see Figs. 4.17 and 4.18). The reflection and transmission coefficients for the differential mode (S_{dd11} and S_{dd21}) and common mode (S_{cc11} and S_{cc21}) of these structures are also shown in these figures. As can be seen, by rotating the resonators 90°, one can selectively choose the suppressed mode in the vicinity of the fundamental resonance frequency.

In Fig. 4.17a, the ELC can be excited only by the magnetic field generated by the line (common mode), but not by the electric field, due to the particle orientation. Similarly, for the MLC of Fig. 4.18a, excitation is only possible under differential mode operation through the electric field generated by the line. The model of the ELC-loaded differential line of Fig. 4.17a is depicted in Fig. 4.19, while that of the MLC-loaded differential line of Fig. 4.18a in Fig. 4.21. It should be noted that the circuit models of ELC- and MLC-loaded differential lines with even- and odd-mode suppression are formally the same as those of CPWs loaded with pairs of SRRs and CSRR-loaded microstrip lines without mixed coupling (Sect. 2.2.4), respectively. It is also worth mentioning that in the MLC-loaded line in Fig. 4.18a, for the common mode, some attenuation occurs as can be clearly observed in Fig. 4.18a, as a consequence of the MLC etching (i.e. a defected ground structure) that disturbs the return current flowing on the ground plane modifying the line parameters. In the circuit model, the resonator is opened, and the resulting model is that of a transmission line with modified shunt capacitance (this does not preclude the fact that the MLC etching can modify the effective inductance of the line as well).

On the other hand, in the structures of Figs. 4.17b and 4.18b, both electric and magnetic couplings between the line and the resonator must be considered. In fact,

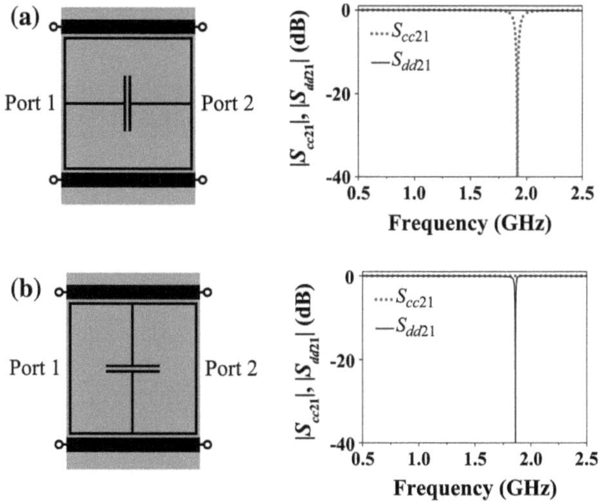

Fig. 4.17 Selective mode suppression in a microstrip differential line loaded with an ELC resonator whose **a** magnetic and **b** electric wall is aligned with the symmetry plane of the line. The substrate is *Rogers RO3010* with $h = 1.27\,\text{mm}$ and $\epsilon_r = 10.2$. Line dimensions are: $W = 1.1\,\text{mm}$ and $S = 10.4\,\text{mm}$ (50-Ω even- and odd-mode characteristic impedance). Resonator dimensions are (see Fig. 2.25): $w_1 = 4\,\text{mm}$, $w_2 = w_3 = l_1 = s = 0.2\,\text{mm}$ and $l_2 = l_3 = 10\,\text{mm}$

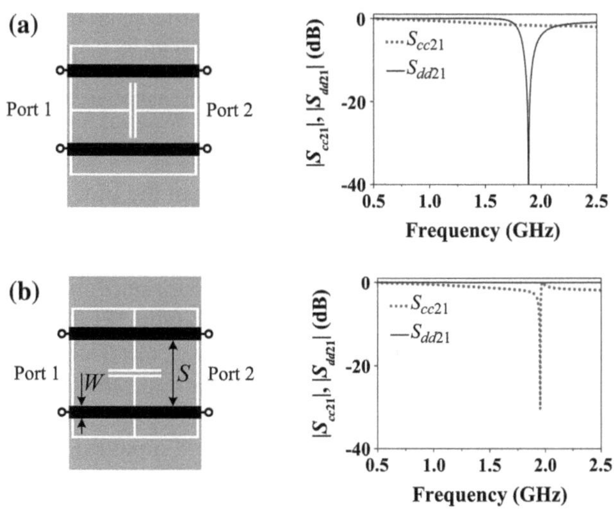

Fig. 4.18 Selective mode suppression in a microstrip differential line loaded with an MLC resonator whose **a** electric and **b** magnetic wall is aligned with the symmetry plane of the line. The substrate is *Rogers RO3010* with $h = 1.27\,\text{mm}$ and $\epsilon_r = 10.2$. Line dimensions are: $W = 1\,\text{mm}$ and $S = 5\,\text{mm}$ (50-Ω even- and odd-mode characteristic impedance). Resonator dimensions are (see Fig. 2.25): $w_1 = 4\,\text{mm}$, $w_2 = w_3 = l_1 = s = 0.2\,\text{mm}$ and $l_2 = l_3 = 10\,\text{mm}$

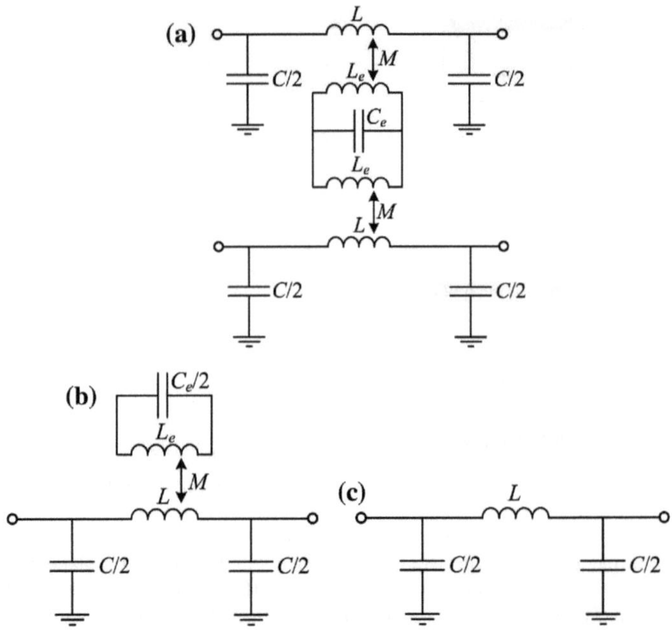

Fig. 4.19 **a** Equivalent circuit model of a microstrip differential line loaded with an ELC whose magnetic wall is aligned with the line, and simplified circuit model for the **b** even and **c** odd mode. The inductance and capacitance of the individual lines are modeled by L and C, respectively. The ELC resonator is modeled by the inductances L_e and the capacitance C_e, and the magnetic coupling between the lines and the resonator is accounted for through the mutual inductances M

the electric coupling is the main driving mechanism of the ELC for the differential mode since the magnetic coupling is not able to induce resonant currents by itself. Analogously, the magnetic coupling between the line and the MLC is absolutely necessary to excite the particle for the common mode, since resonance cannot be driven exclusively through the electric coupling. Therefore, ELC- and MLC-loaded differential lines for odd- and even-mode suppression, respectively, suffer from magnetoelectric coupling that is taken into account in the equivalent circuits, which are represented in Figs. 4.23 and 4.25.

The topology of the ELC and MLC resonators is modified as compared with that of Figs. 4.17 and 4.18. In particular, the aspect ratio w_1/w_2 and l_1 are increased. Thereby the electrical size of the particles at resonance decreases, and accordingly the lumped element circuit models are valid in a wider frequency range. The electrically small elemental cells and the comparison between the circuit and electromagnetic

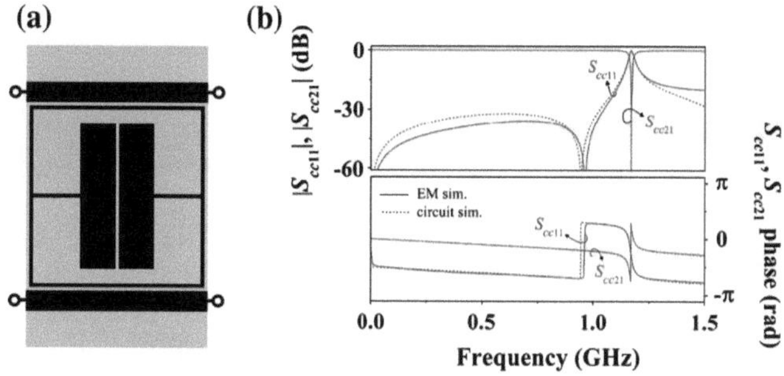

Fig. 4.20 **a** Layout for the ELC-loaded line in Fig. 4.17a with $l_1 = 2\,\text{mm}$ and $w_1 = 8\,\text{mm}$, and **b** common-mode frequency response. The parameters in the circuit of Fig. 4.19b are extracted from the procedure in [10] and assuming an arbitrary M: $L = 4.57\,\text{nH}$, $C = 2.2\,\text{pF}$, $L_e = 4.57\,\text{nH}$, $C_e = 8.14\,\text{pF}$, and $M = 1\,\text{nH}$

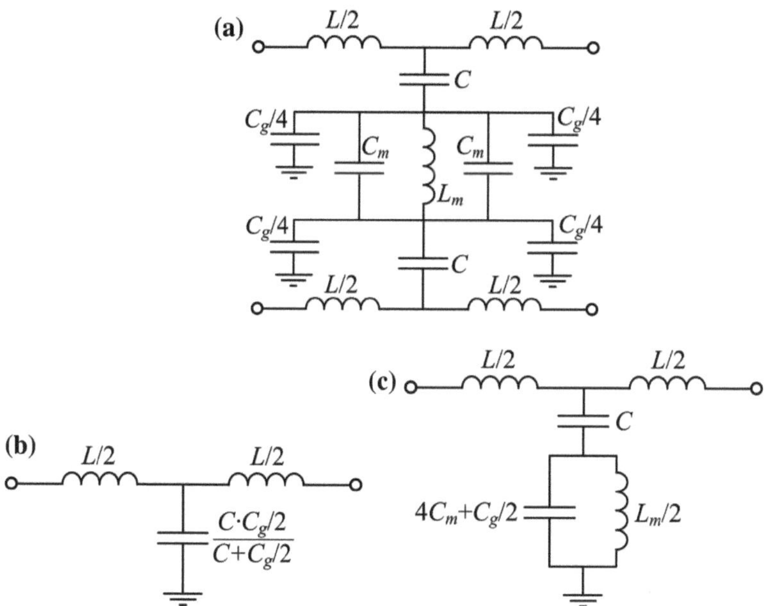

Fig. 4.21 **a** Equivalent circuit model of a microstrip differential line loaded with an MLC whose electric wall is aligned with the line, and simplified circuit model for the **b** even and **c** odd modes. The line inductance and capacitance are modeled, respectively, by L and C, where C acts as the coupling capacitance between the line and the MLC. The resonator is modeled by the inductance L_m, the capacitances C_m, and the capacitance C_g (see Fig. 4.50)

simulations for the four topologies under study are shown in Figs. 4.20, 4.22, 4.24 and 4.26. It can be observed that the frequency response of the ELC- and MLC-loaded differential lines are properly predicted by the proposed equivalent circuits.

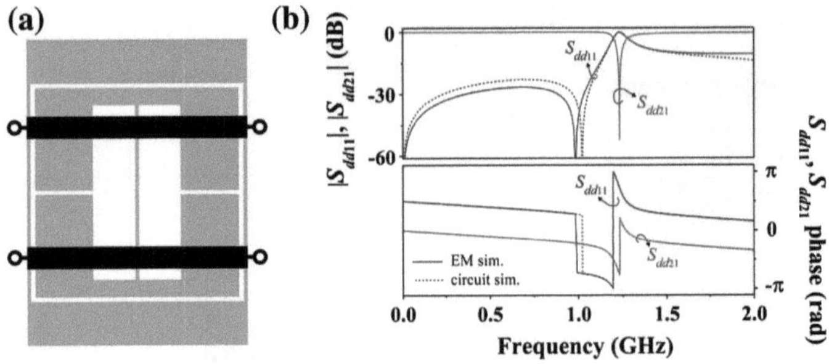

Fig. 4.22 **a** Layout for the MLC-loaded line in Fig. 4.18a with $l_1 = 2$ mm and $w_1 = 8$ mm, and **b** differential-mode frequency response. The circuit parameters in Fig. 4.21c are extracted from the method reported in [9]: $L = 6.07$ nH, $C = 1.53$ pF, $4C_m + C_g/2 = 7.47$ pF, and $L_m = 3.71$ nH

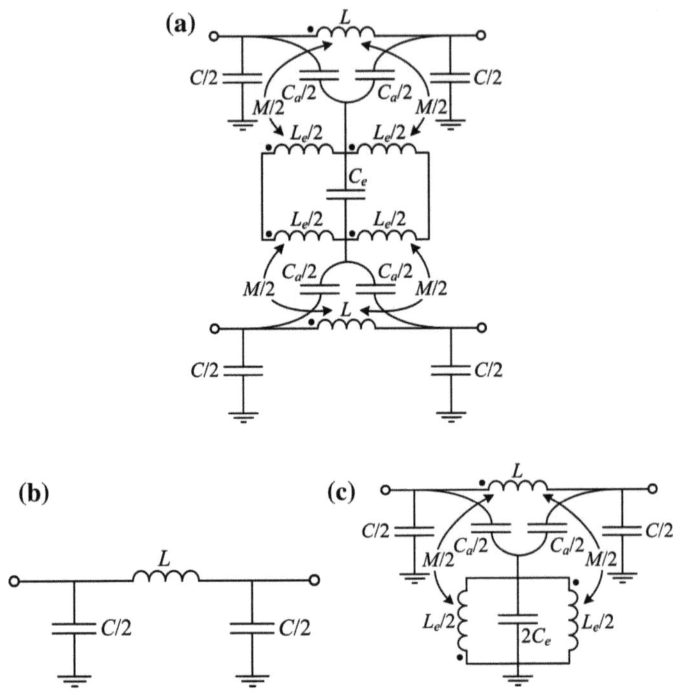

Fig. 4.23 **a** Equivalent circuit model of a microstrip differential line loaded with an ELC whose electric wall is aligned with the line, and simplified circuit model for the **b** even and **c** odd modes. From the circuit in Fig. 4.19a, the resonator is rotated 90° and the capacitances C_a are added standing for the line-to-resonator electric coupling

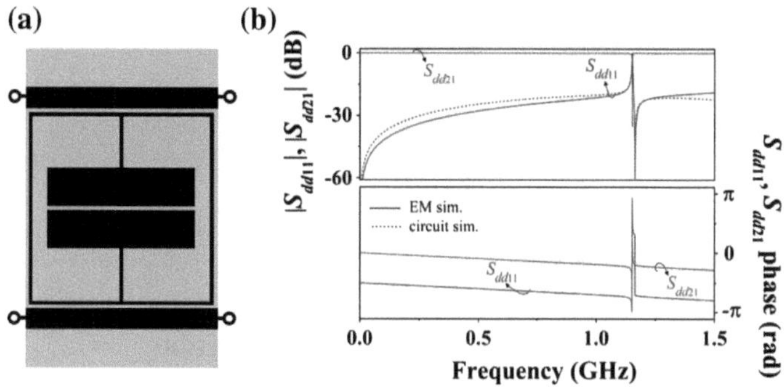

Fig. 4.24 **a** Layout for the ELC-loaded line in Fig. 4.17b with $l_1 = 2\,\text{mm}$ and $w_1 = 8\,\text{mm}$, and **b** differential-mode frequency response. The circuit parameters in Fig. 4.23c are obtained by tuning C_a and readjusting slightly the rest of parameters corresponding to those in Fig. 4.20: $L = 4.09\,\text{nH}$, $C = 1.88\,\text{pF}$, $L_e = 4.62\,\text{nH}$, $C_e = 8.04\,\text{pF}$, $M = 1.12\,\text{nH}$, and $C_a = 0.44\,\text{pF}$

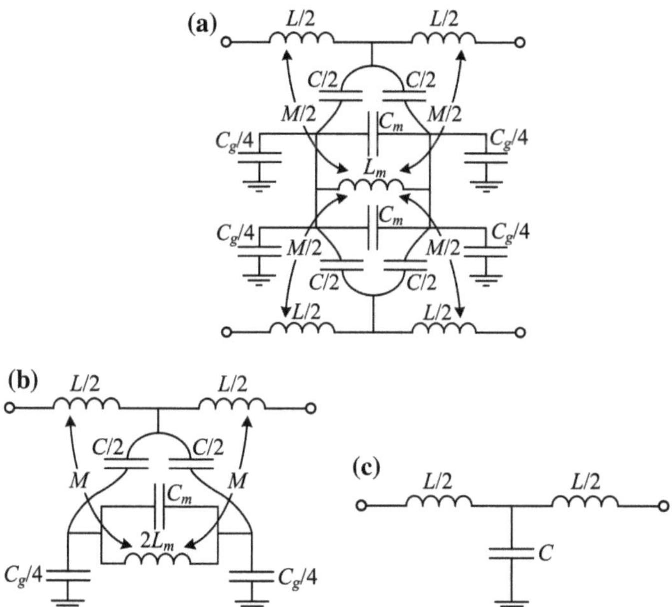

Fig. 4.25 **a** Equivalent circuit model of a microstrip differential line loaded with an MLC whose magnetic wall is aligned with the line, and simplified circuit model for the **b** even and **c** odd modes. From the circuit in Fig. 4.21a, the resonator is rotated 90°, and line-to-resonator magnetic coupling is introduced by the mutual inductances M

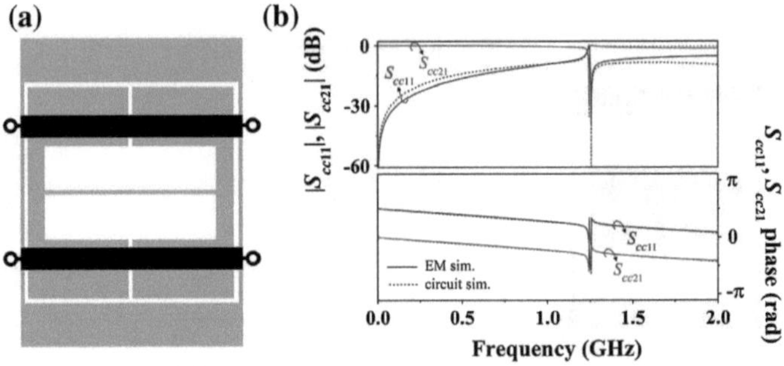

Fig. 4.26 **a** Layout for the MLC-loaded line in Fig. 4.18b with $l_1 = 2$ mm and $w_1 = 8$ mm, and **b** common-mode frequency response. The parameters in Fig. 4.25b are obtained from those in the caption of Fig. 4.22, adding M, and by optimizing the circuit elements to fit the electromagnetic simulation: $L = 9.3$ nH, $C = 1.53$ pF, $L_m = 3.64$ nH, $C_m = 0.93$ pF, $C_g = 7.5$ pF and $M = 0.5$ nH ($4C_m + C_g/2 = 7.47$ pF)

It should be highlighted that due to the bisymmetry of the resonators, all the four configurations and their corresponding circuits are symmetric with regard to the ports.[9] As a result, the reflection coefficients for both the common and differential modes are identical, i.e. $S_{cc11} = S_{cc22}$ and $S_{dd11} = S_{dd22}$. Notice also that the electric and magnetic couplings between the pair of microstrip lines is neglected in the equivalent circuits due to the significant separation between the individual lines (in the case of the ELC resonator, this is necessary to accommodate it). Note also that the considered particles are designated as electric- or magnetic-coupled *LC* resonators following the terminology in the literature [17], which refers to the fact that they can be excited only by either electric or magnetic fields when the fields are uniform (Sect. 2.3). Despite this, in the considered structures, the ELC and MLC resonators are excited by highly non-uniform electromagnetic fields.

Theory is experimentally validated by means of a balanced bandstop filter composed of three cascaded and coupled MLCs with different longitudinal dimensions. As mentioned before, the reported equivalent circuit model of an MLC-loaded differential microstrip line for the odd mode is formally equivalent to that of a CSRR-loaded microstrip line where efficient signal rejection can be obtained. The odd-mode characteristic impedance of the differential microstrip line is not 50 Ω, but lower, since it is designed to mitigate the passband insertion loss due to the presence of the resonators. The filter and its frequency response are shown in Fig. 4.27. The measured response exhibits a rejection of more than 20 dB for differential signals over a fractional bandwidth of 14.3 % centered at 1.3 GHz. This narrow balanced bandstop filter with low insertion loss in the passbands can be useful to suppress undesired differential signals.

---

[9]Note that simultaneous symmetry with regard to both the longitudinal and transverse midplanes cannot be satisfied with resonators exhibiting only a single symmetry plane (e.g. SRRs and CSRRs).

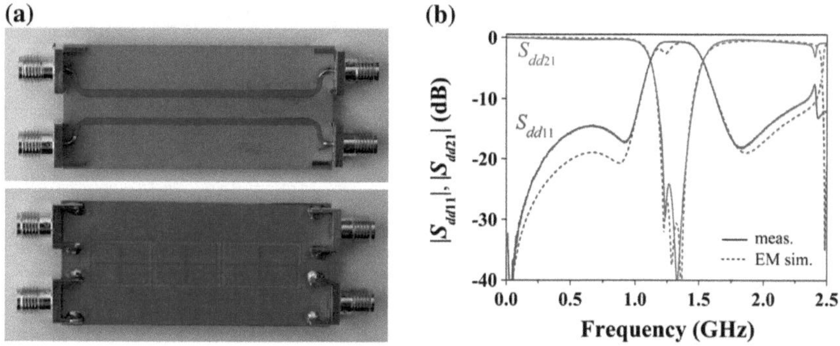

Fig. 4.27 **a** Photograph of top and bottom face, and **b** differential-mode frequency response for the designed differential bandstop filter. The substrate is *Rogers RO3010* with $h = 1.27\,\text{mm}$, $\epsilon_r = 10.2$, and $\tan\delta = 0.0023$. Line dimensions are: $W = 1.6\,\text{mm}$ and $S = 5\,\text{mm}$ (42-Ω odd-mode impedance). MLC dimensions are: $w_1 = 8\,\text{mm}$, $w_2 = l_1 = s = 0.2\,\text{mm}$, $w_3 = 0.8\,\text{mm}$, $l_2 = [13.2; 15.1; 17.6]\,\text{mm}$, and $l_3 = 10\,\text{mm}$. The inter-resonator distance is $d_R = 0.2\,\text{mm}$

4.3 On the Misalignment of Symmetry Planes

This section extends the analysis concerning the resonator excitation explained in the previous section to any resonator orientation. Therefore, it is also considered that the resonator can be misaligned with the line. *Assume that a symmetric transmission line is loaded with a symmetric resonator* (see Fig. 4.28).

- *If the two symmetry planes are aligned, there are two complementary situations* (as explained in Sect. 4.2):
  - *If the symmetry planes are of the same nature, resonance is established, and there is signal suppression in the vicinity of the resonance frequency* ($|S_{21}| \approx 0$).
  - *If the symmetry planes are of distinct nature, resonance is inhibited, and there is no signal suppression* ($|S_{21}| \approx 1$).
- *If the symmetry planes are not aligned, resonance occurs regardless of the nature of the symmetry planes, and signal rejection in the form of a notch arises. The reason is that there cannot be a perfect cancellation of the electromagnetic field components illuminating the resonator, which can be excited electrically and/or magnetically.* Interestingly, *the stopband characteristics (frequency, bandwidth, and/or attenuation) are modulated by the line-to-resonator coupling, which in turn is determined by the degree of misalignment* ($0 < |S_{21}| < 1$).[10]

Therefore, the transmission characteristics of transmission lines loaded with symmetric resonators can be controlled by the orientation of the resonator with regard to the line, which determines the coupling between these two elements.

[10]The phase response is also modulated by the coupling, and may be a subject of future work since it is not considered throughout this thesis.

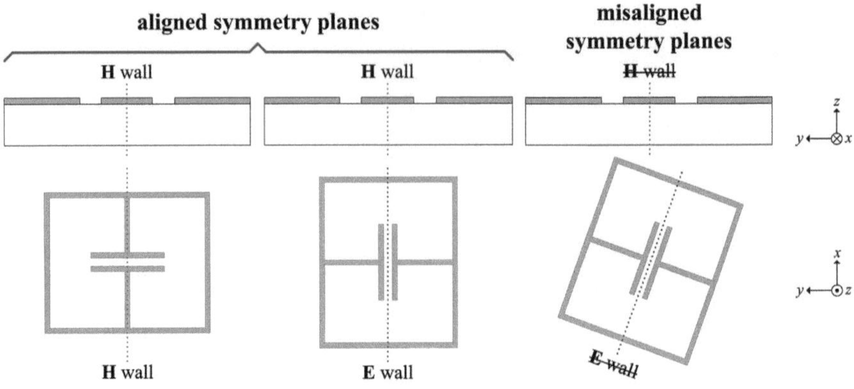

Fig. 4.28 Illustration of stopband modulation by alignment/misalignment of symmetry planes. A CPW loaded with an ELC resonator is considered, where the stopband characteristics may be modulated by rotating the resonator. Evidently, for misaligned loadings, there is no magnetic wall at the symmetry plane of the line

In order to model different orientations, equivalent circuit models with variable parameters depending on the resonator position are required. Consequently, any change in the resonator position should be modeled by adjusting the appropriate circuit parameters. Obviously, when the symmetry planes do not coincide there is neither an electric wall nor a magnetic wall at the symmetry plane of the line, and the even/odd-mode analysis (Appendix C) cannot be applied. It is also important to highlight that to model the notch magnitude, lossy equivalent circuit models are necessary (otherwise, $|S_{21}| = 0$ in the absence of losses). The modeling of losses is specially important when the transmission notch is narrow, as it is for small asymmetries. In this thesis, losses in an equivalent circuit model are evaluated in Sect. 4.3.3 [18].

Any misalignment can be straightforwardly accounted for by relative linear (through a straight line) and/or angular (rotation) displacements between the line and the resonator. The misalignment of symmetry planes was proposed as a novel sensing mechanism for alignment- and displacement-oriented sensing for the first time in [19]. Furthermore, in this thesis, the misalignment of symmetry planes has been used not only as a sensing mechanism, but also as an encoding mechanism. The design of microwave sensors and radio-frequency (RF) barcodes by that means is addressed in Chaps. 6 and 7, respectively.

Next subsections present the results for the considered resonator-loaded transmission lines where the resonator is aligned and misaligned with the line. As will be shown, the modulation (controllability) of the stopband in terms of alignment/misalignment is validated by variable electric or magnetic line-to-resonator couplings.

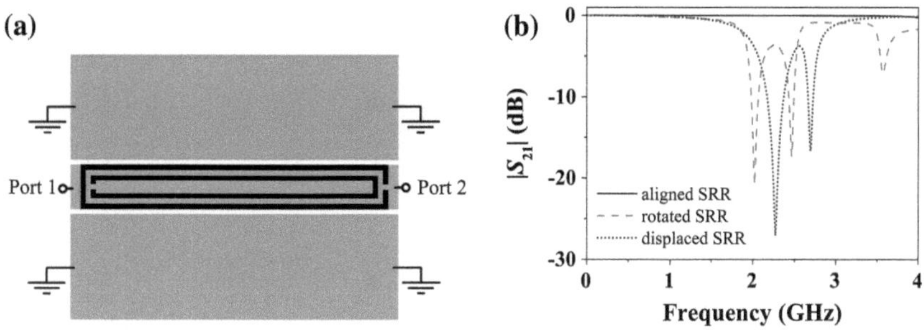

Fig. 4.29 **a** CPW symmetrically loaded with a rectangular-shaped SRR, and **b** simulated transmission coefficient for this aligned case and for the cases where the SRR is either displaced by 0.5 mm or rotated from its center by 10°. Dimensions are (see Fig. 4.6): $W = 1.67$ mm and $G = 0.2$ mm (50-Ω line), $c = d = 0.2$ mm, $l_1 = 12$ mm and $l_2 = W = 1.67$ mm. The substrate is *Rogers RO3010* with thickness $h = 127\,\mu$m, relative permittivity $\epsilon_r = 10.2$, and loss tangent $\tan\delta = 0.0023$. Reprinted with permission from [19]

4.3.1 SRR- and FSIR-Loaded Coplanar Waveguides

Naqui et al. reported for the first time the stopband modulation functionality by means of misalignments in [19]. As shown in Fig. 4.29, an SRR is etched symmetrically on the back side of the substrate in a CPW. In this figure, the transmission coefficient for this symmetric configuration is plotted, as well as that for the structures that result by displacing laterally or by rotating the resonator from the aligned configuration. As expected, the results reveal that if the SRR is symmetrically etched, the structure is transparent at SRR resonance. The SRR is not excited if the slits are perfectly aligned with the symmetry plane of the CPW because the symmetry plane of the SRR is an electric wall at resonance, but the symmetry plane of a CPW structure is a magnetic wall for the CPW (even) mode. However, if the symmetry is altered by simply displacing or rotating the SRR, a notch appears in the transmission coefficient at the resonance frequency of the SRR. There are actually visible two nearby notches which correspond to the inner and outer rings. These two resonances are close to each other because the rings have similar lengths, and also owing to the fact that are loosely coupled as will be explained shortly. Due to the resonator orientation, line-to-resonator magnetoelectric coupling is present. In this seminal work, however, an equivalent circuit model is not presented.

The effects of substrate thickness are analyzed from Fig. 4.30, where two different thicknesses are considered. One of the effects for thin substrates is that the coupling between the two split rings forming the SRR is seen to be extremely limited at the expense of an increase in the coupling between the rings and the CPW (the substrate thickness, 127 μm, is indeed smaller than the inter-ring separation, 200 μm). As a result, the SRR fundamental resonance frequency is identical to the one obtained with the larger ring in the absence of the smaller one. By contrast, for thicker substrates, the lowest two resonances of the SRR experience a frequency shift (the fundamental

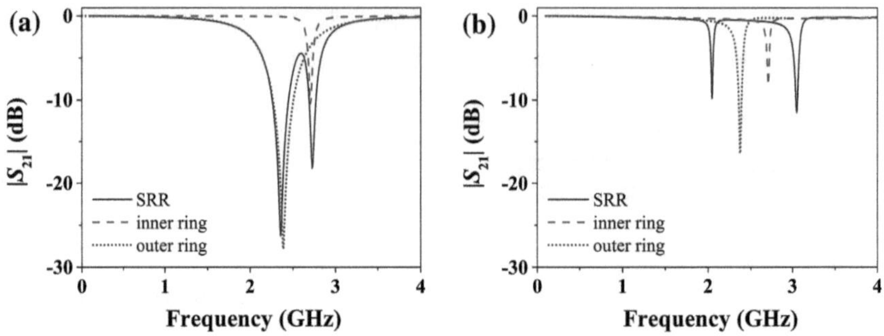

Fig. 4.30 Simulated transmission coefficient of the structure in Fig. 4.29 by laterally displacing 0.5 mm the SRR, and replacing the SRR with only the inner or outer ring. **a** $h = 127\,\mu$m and **b** $h = 635\,\mu$m (38-Ω line). Reprinted with permission from [19]

resonance shifts downward and the second one upward) caused by an evident inter-ring coupling. Another effect is that the line-to-resonator coupling enhances as the substrate thickness decreases. Thereby deeper notches are obtained, and smaller misalignments can be detected. Thus, thin substrates are preferred, even though this can result in a negligible inter-ring coupling. Therefore, the SRR structure does not behave as a pair of coupled split ring resonators (Sect. 2.3.1), and one cannot take advantage of its supposedly electrically small size.[11] For such a reason, single split-ring resonators can be used indistinguishably, as corroborated in [20] (Sect. 6.4.1).

As an alternative proposal to the SRR and the single split-ring resonator, [21] investigates CPWs loaded with folded stepped-impedance resonators (FSIRs) in terms of alignment/misalignment. The considered topology and its transmission coefficient conditioned to different FSIR positions are shown in Figs. 4.31b and 4.32, respectively. As expected, the more displacement, the stronger the notch (obviously, up to a certain displacement). Hence, the results agree with the theory behind the alignment/misalignment of symmetry planes.

Like in SRR-based CPWs, in FSIR-loaded CPWs there are mixed coupling effects. Despite this, since the considered substrate is thin, by designing the open-ended sections of the FSIRs with a wide width, strong line-to-resonator electric coupling may be obtained, this coupling becoming the dominant one. An equivalent circuit model neglecting the magnetic coupling is presented, which is represented in Fig. 4.33a. Note that although the structure is not symmetric in regard to the ports, the circuit model is symmetric because the magnetic coupling is neglected. In such an equivalent circuit, the electric coupling is modeled by capacitances that depend on the relative position of the resonator. Thus, any symmetry disruption can be modeled by adjusting these capacitances. If the structure is symmetric, $C_{sc1} = C_{sc2} = C_{sc}$ and $C_{sg1} = C_{sg2} = C_{sg}$, so that $C_{sc1} - C_{sc2} = 0$ and $C_{sg1} - C_{sg2} = 0$ (the difference in capacitances related to the individual halves is null). As a result, the resonator

[11] Although the first resonance frequency is no longer quasistatic and the SRR cannot be considered electrically small at that frequency, the symmetry properties of the loaded line are exactly the same.

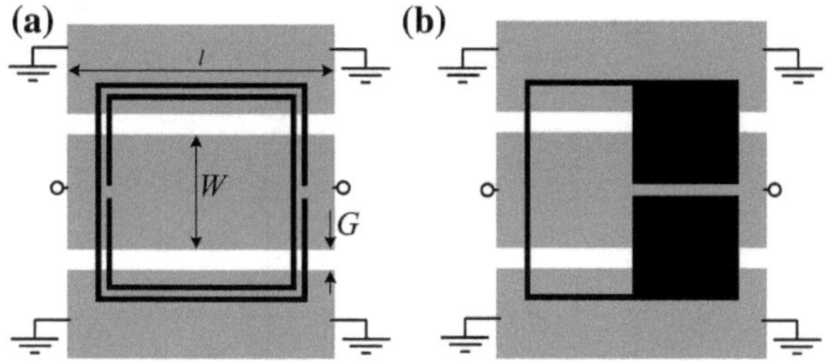

Fig. 4.31 CPW loaded with a symmetrically etched **a** SRR and **b** FSIR. The substrate is *Rogers RO3010* with $h = 0.635\,\text{mm}$, $\epsilon_r = 10.2$, and $\tan\delta = 0.0023$. CPW dimensions are: $l = 9.6\,\text{mm}$, $W = 4\,\text{mm}$ and $G = 0.7\,\text{mm}$ (50-Ω line). SRR dimensions are (see Fig. 4.6): $l_1 = l_2 = 7.6\,\text{mm}$ and $c = d = 0.2\,\text{mm}$. FSIR dimensions are (see Fig. 2.23): $l_1 = l_2 = 7.6\,\text{mm}$, $w_1 = 3.8\,\text{mm}$, $w_2 = 0.2\,\text{mm}$ and $S = 0.4\,\text{mm}$

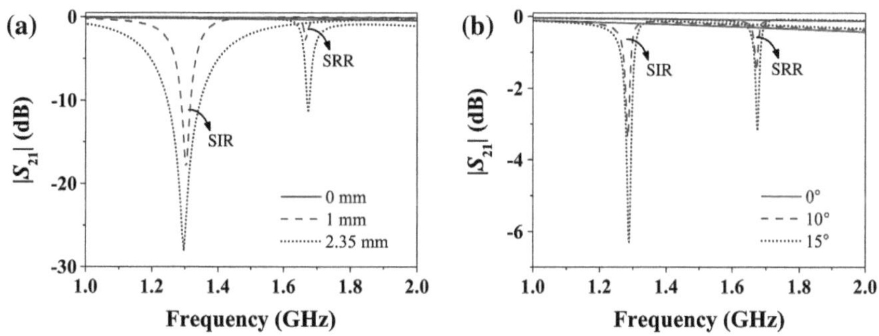

Fig. 4.32 Simulated transmission coefficient of the structures in Fig. 4.31 and that corresponding to the same structures with the resonators **a** laterally displaced or **b** rotated. Reprinted with permission from [21]

$L_s C_s$ is opened by symmetry in the shunt branch, and the resulting model is that of a conventional transmission line, as depicted in Fig. 4.33b. By contrast, if symmetry is broken, the capacitances are unbalanced and resonance can be established. In the simplest case in which $C_{sc2} = C_{sg1} = 0$ (or $C_{sc1} = C_{sg2} = 0$), the resulting circuit is that in Fig. 4.33c. The transmission zero is determined by the resonator formed by the FSIR parameters and the capacitances C_{sci} and C_{sgi}, rather than being given by the intrinsic resonance of the FSIR.[12] By neglecting the capacitance C_s, the model is equivalent to that of a transmission line loaded with a shunt series resonator. These circuit models are verified in Fig. 4.34 for linear displacements, where the circuit simulations are in good agreement with the electromagnetic ones. As can be noticed, the simplified model (without the magnetic coupling) suffices to accurately describe the transmission coefficient magnitude (with the exception of the notch depth, since the circuit model is assumed to be lossless).

[12] Due to the capacitive coupling, like in a CSRR-loaded line, the transmission zero is not given by the intrinsic resonance of the loading resonator.

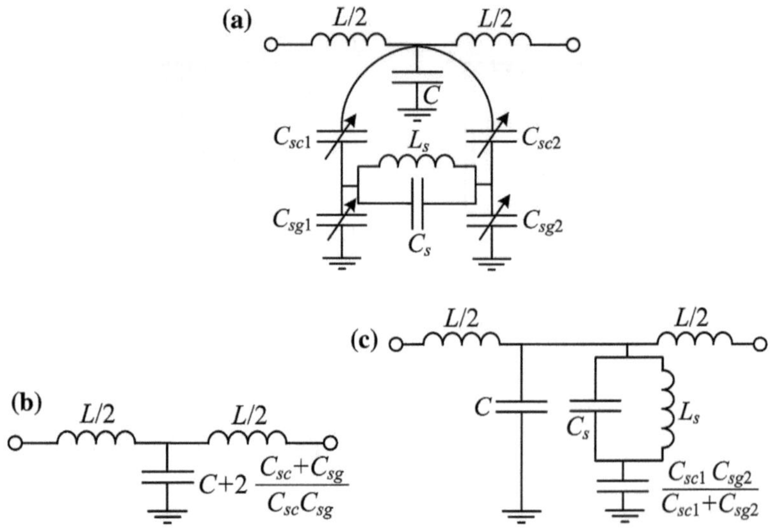

Fig. 4.33 **a** Equivalent circuit model of a CPW loaded with an FSIR, neglecting the line-to-resonator magnetic coupling. **b** Simplified model when the FSIR is symmetrically loaded to the line. **c** Simplified model when the FSIR is laterally displaced such that $C_{sc2} = C_{sg1} = 0$. The line inductance is L, the line capacitance is C, the resonant tank L_sC_s models the FSIR inductance-capacitance, C_{sci} is the capacitance between the central strip and one of the halves (i indicating the half) of the FSIR, and C_{sgi} is the capacitance between each FSIR half and each lateral ground plane

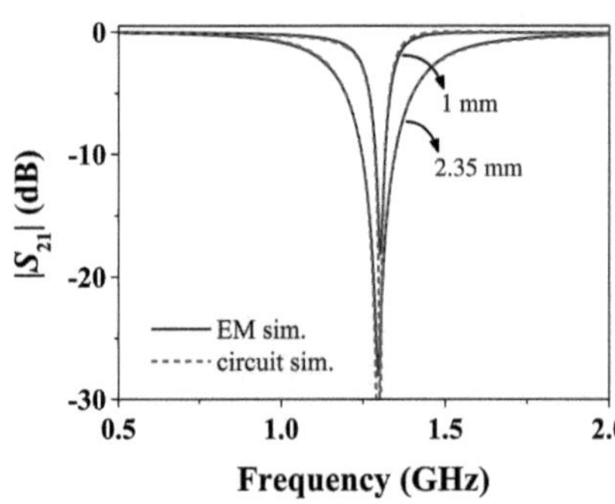

Fig. 4.34 Simulated transmission coefficient of the FSIR-loaded CPW in Fig. 4.31 with the resonator laterally displaced 1 and 2.35 mm. The circuit elements are extracted as follows. (*i*) The line elements are estimated as the inductance and capacitance of the CPW section; $L = 2.4$ nH and $C = 0.97$ pF. (*ii*) The FSIR elements and the line-to-resonator capacitive coupling are found by curve fitting the notch frequency and bandwidth. The FSIR elements are: $L_s = 13.8$ nH and $C_s = 0$ pF. The variable coupling capacitances are: for 1 mm, $C_{sc1} = 0.57$ pF, $C_{sc2} = 1.8$ pF, $C_{sg1} = 1.6$ pF, and $C_{sg2} = 0.4$ pF; for 2.35 mm, $C_{sc1} = C_{sg2} = 0$ pF, $C_{sc2} = 2.2$ pF and $C_{sg1} = 2.2$ pF. Reprinted with permission from [21]

Moreover, in [21] CPWs loaded with either SRRs or FSIRs with the same dimensions (see Fig. 4.31) are also compared. The results, plotted in Fig. 4.32, indicate that lower notch frequencies are obtained using FSIRs. As mentioned before, the inter-ring coupling in SRRs is limited to a considerable extent, preventing from their quasistatic resonance. By contrast, very low resonance frequencies may be obtained utilizing FSIRs due to the significant achievable broadside electric coupling, which directly influences the transmission zero frequency. Accordingly, for thin substrates, FSIR-loaded CPWs are electrically smaller at resonance than those based on SRRs. Moreover, using FSIRs not only smaller notch frequencies are obtained, but also wider notches. This is also attributed to the strong electric coupling in FSIR-based CPWs.

Measured results that prove the previous theory are reported in Sects. 6.3.1 and 6.4 within the context of microwave spatial sensors.

4.3.2 SIR-Loaded Microstrip Lines

Naqui and Martín analyze a microstrip line loaded with an SIR in [22]. Note that it is not possible to somehow load symmetrically a conventional microstrip line with a strip resonator (without connecting the resonator to the line), and accordingly three metal layers are required. It is assumed that the SIR (folded or unfolded) is etched on an additional layer located above the microstrip line, as illustrated in Fig. 4.35a. Since the symmetry plane of an SIR behaves as an electric wall, if an SIR is aligned with the line, the structure is transparent to signal propagation.[13] On the contrary, by rupturing symmetry through a displacement of the resonator (along the transverse direction), a notch appears.

An unfolded SIR is bisymmetric and does not suffer from cross-polarization effects. Thus, with a view to deriving an equivalent circuit model, the unfolded configuration of the SIR is considered in order to avoid mixed coupling, as is shown in Fig. 4.35b. As a result, the SIR is electrically coupled to the microstrip line. Using the unfolded SIR, in addition, the structure is symmetric when looking into the ports. An equivalent circuit model for the SIR laterally displaced is reported, which is depicted in Fig. 4.36. In this model, a variable capacitance, C_s, accounts for the transverse displacement modeling the coupling between the signal strip and the resonator and the resonator-to-ground coupling as well. The circuit model validation is plotted in Fig. 4.37. The principle of operation is also supported by experimental data, provided in Sect. 6.3.2.

[13]Although the unfolded SIR is bisymmetric, it is considered only the symmetry plane with an electric wall that bisects the resonator like in the folded configuration.

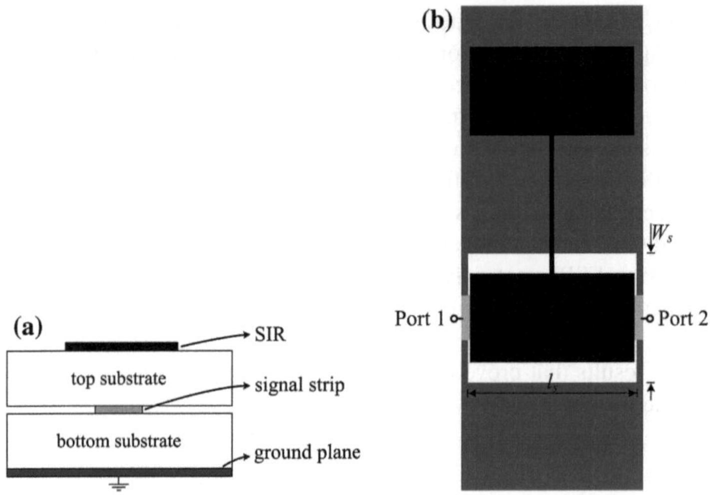

Fig. 4.35 **a** Cross-sectional view of a microstrip line loaded with an SIR, etched on top of it. **b** SIR-loaded microstrip line for a displacement corresponding to the strongest notch. A window in the ground plane is etched to improve the performance working as a displacement sensor, as will be explained in Sect. 6.3.2. The dimensions are: the microstrip line width is 2 mm; for the slotted window, $W_s = 5.8$ mm and $l_s = 7.8$ mm; for the SIR, the wide and narrow widths are 7.6 mm and 0.2 mm, and their lengths are 4 mm and 6 mm, respectively. The substrates are *Rogers RO3010* with $h = 0.254$ mm and $\epsilon_r = 11.2$

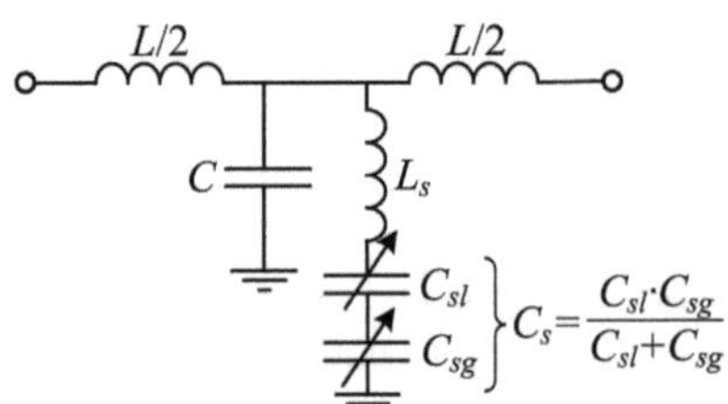

Fig. 4.36 Equivalent circuit model of an SIR-loaded microstrip line for a displacement corresponding to the strongest notch like the one shown in Fig. 4.35. The inductance and capacitance of the line are L and C, the inductance of the SIR is L_s, the SIR-to-line capacitance is C_{sl}, and the SIR-to-ground capacitance is C_{sg}

4.3.3 *ELC-Loaded Coplanar Waveguides*

Naqui et al. load a CPW with an ELC resonator whose axes intersect perpendicularly to each other, and a dynamic loading by rotating the resonator is considered [23]. It is shown that the transmission coefficient of the structure depends on the angular orientation of the resonator. Since an ELC resonator has two symmetry planes, one being an electric wall and the other one a magnetic wall, the resonator can load symmetrically a CPW by aligning the line axis with either its electric or magnetic wall.

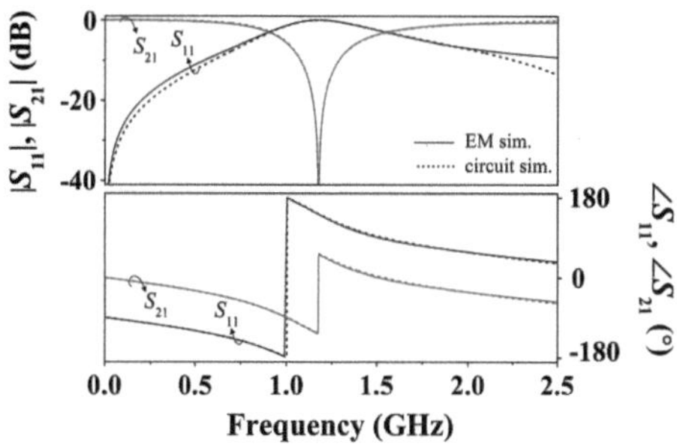

Fig. 4.37 Transmission and reflection coefficients obtained from the lossless electromagnetic and circuit simulations for the SIR-loaded microstrip line in Fig. 4.35. The circuit parameters are extracted by adapting the procedure reported in [9] to the specific considered circuit; $L = 4.43$ nH, $C = 1.58$ pF, $L_s = 6.90$ nH, and $C_s = 2.66$ pF. Reprinted with permission from [22]

These two orientations are illustrated in Fig. 4.38, and the corresponding frequency responses are shown in Fig. 4.39. For the fundamental CPW mode, an ELC resonator cannot be excited if its electric wall is aligned with the line axis, and the loaded line is transparent. Conversely, the resonator is driven if its magnetic wall is aligned with the line, and a zero in the transmission coefficient appears. For intermediate (asymmetric) orientations, the resonator is also excited and the notched response (notch bandwidth and depth) is dependent on the resonator orientation. Obviously, by symmetry and owing to the orthogonal electric and magnetic walls of the resonator, the angular orientation range is limited to 90°. The optimum orientation for the strongest attenuation is the one corresponding to a perfect overlapping between the ELC magnetic wall and the symmetry plane of the line. Therefore, by spatially rotating the resonator the frequency response switches from an all-pass to a stopband behavior and vice versa, the attenuation being dependent on the angular orientation.

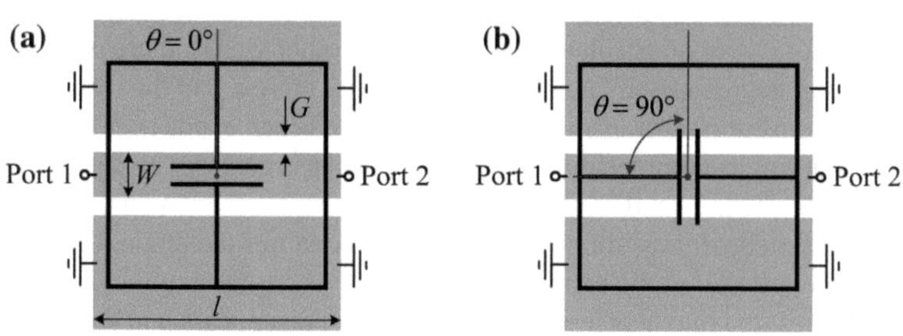

Fig. 4.38 CPW symmetrically loaded with an ELC resonator etched on the back side of the substrate. The orientation between the CPW and the resonator is determined by the rotation angle θ; the line axis is perfectly aligned **a** with the resonator electric wall for $\theta = 0°$, and **b** with the resonator magnetic wall for $\theta = 90°$. CPW dimensions are: $W = 0.78$ mm and $G = 2$ mm (50-Ω line). In reference to Fig. 2.25, ELC dimensions are: $w_1 = 4$ mm, $w_2 = w_3 = l_1 = s = 0.2$ mm, and $l_2 = l_3 = 10$ mm. The substrate is *Rogers RO3010* with $h = 1.27$ mm and $\epsilon_r = 11.2$

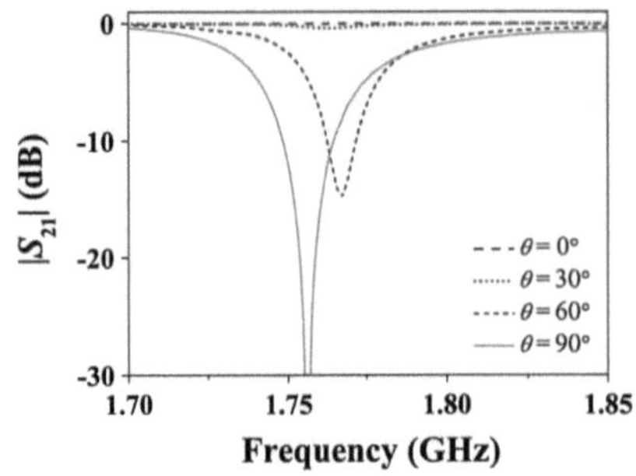

Fig. 4.39 Lossless transmission coefficient of the structure in Fig. 4.38 with different resonator rotation angles. The angles $\theta = 0°$ and $\theta = 90°$ correspond to the extreme cases where the symmetry plane of the line (a magnetic wall) is perfectly aligned with the resonator electric wall and the magnetic wall, respectively. Reprinted with permission from [23]

Note that despite the nomenclature to designate the resonator, in these ELC-loaded CPWs the resonator is excited by the non-uniform fields generated by the transmission line. Interestingly, the resulting line-to-resonator coupling is not electric, but magnetic.[14] The magnetic field lines penetrating the individual loops of the resonator are in opposite directions to each other, and this makes the resonance establishment possible. Specifically, when the electric wall of the resonator is aligned with the line ($\theta = 0°$), there is a total cancellation between the inward and outward magnetic field components axial to each loop. As the resonator is rotated, the cancellation gradually decreases until it finally vanishes when the magnetic wall of the ELC resonator is aligned with the line axis ($\theta = 90°$).

An equivalent circuit model valid for any arbitrary loading angle, depicted in Fig. 4.40a, is proposed in the expanded work of [23], namely in [18]. The interpretation of the model in terms of the mutual inductances is as follows. When $\theta = 0°$, due to symmetry, $M_\theta = M^\theta \neq 0$, which means that the currents flowing on the CPW induce a pair of equal and anti-phase voltages in the loops; no net voltage is induced in the resonator due to an absolute cancellation. In other words, there is no net magnetic coupling, the resonator cannot be magnetically driven, and the resulting model is that of a conventional transmission line. On the other hand, as θ increases the magnetic coupling is complementarily distributed; M^θ increases at the expense of a decrease in M_θ. Hence, for $\theta > 0°$, a net induced voltage arises and the line is indeed capable of magnetically exciting the resonator. The larger the angle, the higher the induced voltage. Thus, at the upper limit, $\theta = 90°$, M^θ is maximum while M_θ completely vanishes; the magnetic coupling cancellation disappears and the resonator is tightly coupled to the line. The resulting simplified equivalent circuits for the two

[14]This was also seen in common-mode suppressed differential microstrip lines by means of ELC resonators (Sect. 4.2.3).

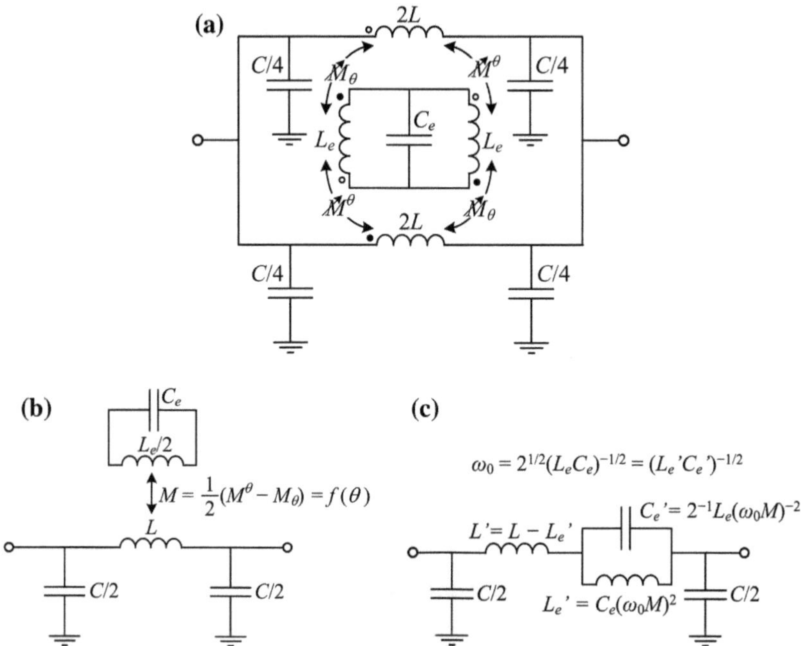

Fig. 4.40 **a** Lossless equivalent circuit model of a CPW loaded with an ELC resonator. **b** Simplified circuit model. **c** Transformed simplified circuit model. The ELC resonator is represented by the inductances L_e and the capacitance C_e, modeling the inductive loops and the capacitive gap, respectively. The CPW is modeled by its inductance and capacitance, L and C, respectively, and it is divided into the two halves with respect to its symmetry plane. The line-to-resonator coupling is modeled through the angle-dependent mutual inductances M_θ and M^θ between each half and each loop (different dots are used to distinguish the different magnetic couplings associated to each of the CPW halves)

extreme orientations are depicted in Fig. 4.41, and the geometrical division of the mutual inductances is illustrated in Fig. 4.42.

In summary, the frequency response of the proposed circuit depends directly on the line-to-resonator magnetic coupling, which in turn is determined by the resonator angular orientation, θ. Even though the physical understanding of magnetic coupling cancellation relative to the resonator orientation is well illustrated by the circuit model in Fig. 4.40a, that circuit is clearly somewhat complex and has too many parameters for both parameter extraction and design purposes. An equivalent and simplified circuit model is the one depicted in Fig. 4.40b, where an effective (or net) mutual inductance M is defined, and the transmission zero frequency is found to be

$$\omega_0 = \frac{1}{\sqrt{L_eC_e/2}}. \tag{4.1}$$

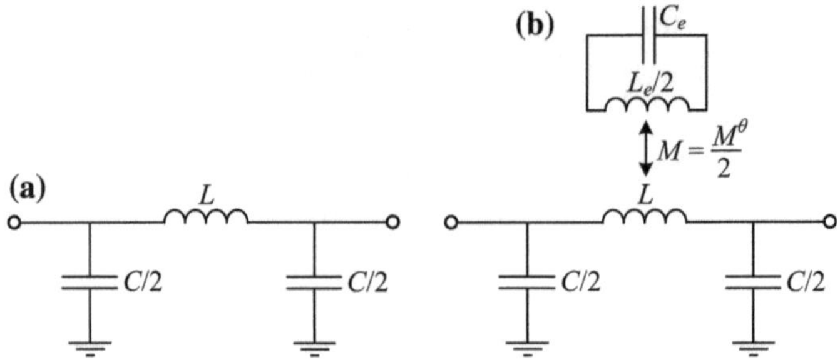

Fig. 4.41 Lossless equivalent circuit model of a CPW loaded with an ELC resonator for the two extreme symmetric orientations. **a** $\theta = 0°$ where $A_\theta = A^\theta = A_l/2$, $M_\theta = M^\theta \propto A_l/2$, $A = 0$, and $M = 0$. **b** $\theta = 90°$ where $A_\theta = 0$, $A^\theta = A_l$, $M_\theta = 0$, $M^\theta \propto A_l$, $A = A_l$, and $M \propto A_l$

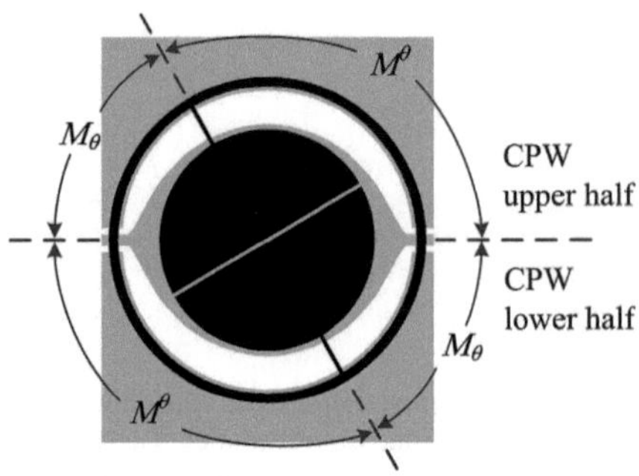

Fig. 4.42 Layout of a circularly-shaped structure with a large resonator capacitor (for a specific angle of 30°) where the geometrical division associated to the mutual inductances M_θ and M^θ is illustrated. The considered substrate is *Rogers RO3010* with $h = 1.27$ mm, $\epsilon_r = 11.2$, and $\tan\delta = 0.0023$. The dimensions are: for the line, W and G are tapered such that the characteristic impedance is 50 Ω, where the radial distance from the center of the CPW to the lateral ground planes is 7.6 m; for the ELC resonator: mean radius $r_0 = 8.05$ mm, capacitor outer radius $r_1 = 5.6$ mm, $w_2 = s = 0.2$ mm, and $w_3 = 0.5$ mm

Such a model is equivalent in form to that of a CPW loaded with pairs of SRRs magnetically coupled to it [24–26], which can be transformed to the circuit model in Fig. 4.40c [26–28].

In order to validate the equivalent circuits of Fig. 4.40, a sub-wavelength dimension of the unit cell is required. To this end, the structure shown in Fig. 4.42 is considered which exhibits a large ELC capacitance. Such a structure, in addition, is circularly-shaped with a view to dealing with dynamic angular orientations. By this means, the cell length l is independent on the orientation, and the mutual inductance as a function of the orientation can be easily obtained. The results of the simulations for different resonator orientations are plotted in Fig. 4.43. Good agreement between circuit and electromagnetic simulations in the vicinity of the transmission zero

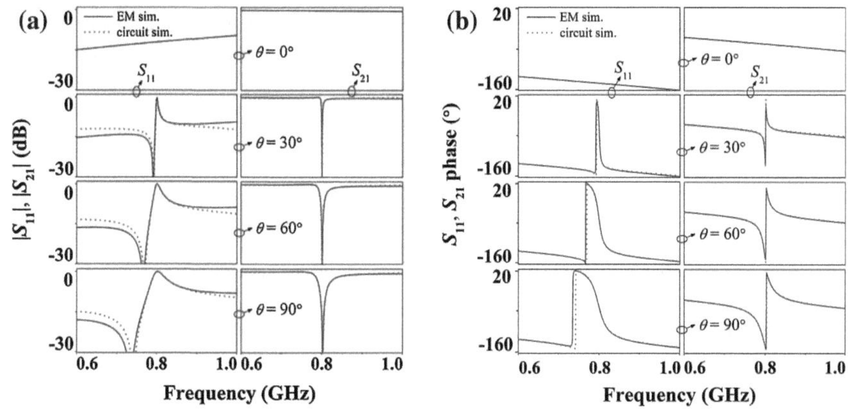

Fig. 4.43 **a** Magnitude and **b** phase of the reflection and transmission coefficients given by lossless electromagnetic simulation for the structure in Fig. 4.42 and by circuit simulation for the models in Fig. 4.40. The circuit parameters are those listed in Table 4.1. Reprinted with permission from [18]

frequency (the region of interest) results, an indicator of validity of the proposed circuits with arbitrary ELC orientation.[15]

It can be noticed from the extracted circuit parameters (Table 4.1) that, with the exception of the mutual inductance, all the circuit parameters are roughly invariant with the rotation angle. According to (4.1), since the parameters L_e and C_e are almost invariant with the angular orientation, the notch frequency is roughly constant with the angle (the maximum frequency deviation is 0.48 %).

It is obvious that L_e is the same for all the rotations because it is assumed to be the inductance of the isolated resonator. One of the reasons to assume that is because the extracted inductance can be compared to analytical expressions available in the literature.[16] Another important reason is that the procedure to extract the inductance is applied once; by contrast, in the case where the inductance (in the presence of the CPW) depends on the line-to-resonator orientation, the procedure should be applied

[15]The circular shape has the disadvantage of increasing the effective electrical length (since the path for wave propagation between the input and output ports increases). As a result, the frequency region where the model accurately predicts the frequency response is not as wide as in a uniform CPW-based structure. Nevertheless, the accuracy is reasonable in the vicinity of resonance.

[16]The technique to estimate the inductance of the resonator, described in Table 4.1, is proposed for the first time in [18]. Interestingly, it can be performed regardless of the loop shape, making the technique very versatile. To validate its usefulness, some results given by this technique are compared with the analytical expressions reported in [27, 30] for circular SRRs, and very close values between the two approaches are obtained. For instance, for a loop with mean radius $r_0 = 8.05$ mm and ring width $c = 0.5$ mm, the extracted input inductance of a single ring is 42.5 nH, while the analytically estimated inductance is 44.03 nH [18].

Table 4.1 Extracted lumped element values of the circuit of Fig. 4.40b for the structure in Fig. 4.42

θ (degrees)	C (pF)	L (nH)	C_e (pF)	L_e (nH)	M (nH)
30	5.86	5.95	3.09	25.6	0.94
60	5.73	6.22	3.09	25.6	1.91
90	5.59	6.40	3.07	25.6	2.72

The circuit parameter extraction method is: (*i*) the elements in Fig. 4.40c are extracted from the electromagnetic simulation following the procedure detailed in [10]; (*ii*) L_e is estimated by means of the procedure proposed in [18], in which the inductance seen looking into the terminals of the isolated loops (i.e. considering neither the resonator capacitor nor the CPW) is extracted from electromagnetic simulation, by virtue of the fact that the input impedance of an electrically small loop is primarily inductive [29]; (*iii*) next, the parameters in Fig. 4.40b are obtained using the indicated transformation equations

for each orientation. Therefore, by considering L_e as the inductance of the isolated resonator, the electromagnetic interaction (coupling) between the CPW and the ELC resonator is merely accounted for by the mutual inductance M.

Regarding the electric coupling, electric interaction between the resonator and the CPW may actually exist, changing the effective capacitances of them. Therefore, C is the effective line capacitance in the presence of the resonator and C_e is the effective capacitance of the resonator in the presence of the CPW, but the electric coupling is not explicitly model because it cannot excite the fundamental resonance of the ELC resonator.

To gain more insight into the mutual inductances, let us find analytical expressions from the definition of the mutual inductance between two circuits, namely [31]

$$M_{ij} = \frac{\Phi_{ij}}{I_j}, \tag{4.2}$$

where Φ_{ij} is the magnetic flux through circuit i generated by a current I_j flowing on a circuit j. The magnetic flux through any surface S can be calculated as the surface integral

$$\Phi_{ij} = \int_S B_n \cdot ds, \tag{4.3}$$

where B_n is the normal component to S of the magnetic flux density generated on the circuit j. As is well known, in the event that B_n is uniform over S, the flux is proportional to the surface area, A_S, that is

$$\Phi_{ij} = B_n A_S. \tag{4.4}$$

In the considered circular ELC resonator, the area of each individual loop is

$$A_l = \frac{\pi[(r_0 - w_3/2)^2 - r_1{}^2]}{2} = A_\theta + A^\theta. \tag{4.5}$$

In reference to Fig. 4.42, the resulting complementary areas are

$$A_\theta = A_l\left(\frac{1}{2} - \frac{\theta}{\pi}\right) = A_l - A^\theta, \tag{4.6a}$$

$$A^\theta = A_l\left(\frac{1}{2} + \frac{\theta}{\pi}\right) = A_l - A_\theta, \tag{4.6b}$$

which are linearly dependent on θ. For an electrically small circular-tapered CPW (as the one considered), the magnetic flux density can be assumed (as a first-order approximation) to be uniform along θ over the areas (4.6). Accordingly, the complementary mutual inductances satisfy the following proportionalities

$$M_\theta = \frac{B_n A_\theta}{I_{2L}} \propto A_l\left(\frac{1}{2} - \frac{\theta}{\pi}\right), \tag{4.7a}$$

$$M^\theta = \frac{B_n A^\theta}{I_{2L}} \propto A_l\left(\frac{1}{2} + \frac{\theta}{\pi}\right), \tag{4.7b}$$

where I_{2L} is the current flowing on the inductance $2L$ modeling half of the CPW. In terms of the effective mutual inductance, since the effective area is

$$A = A^\theta - A_\theta = A_l\left(\frac{2\theta}{\pi}\right), \tag{4.8}$$

and the current flowing on the circuit in this case is that flowing through L, that is

$$I_L = 2I_{2L}, \tag{4.9}$$

the resulting net mutual inductance is[17]

$$M = \frac{B_n A}{I_L} = \frac{1}{2}(M^\theta - M_\theta) \propto A_l\left(\frac{2\theta}{\pi}\right), \tag{4.10}$$

satisfying

$$M_\theta = M\left(\frac{\pi}{2\theta} - 1\right) = M^\theta - 2M, \tag{4.11a}$$

$$M^\theta = M\left(\frac{\pi}{2\theta} + 1\right) = M_\theta + 2M. \tag{4.11b}$$

Expression (4.10) clearly states not only that the net mutual inductance modeling the line-to-resonator magnetic coupling is sensitive to θ, but also that it depends (in a first-order approximation) linearly on θ. It should also be noted that the total

[17] In the limiting case where $\theta = \pi/2$ and $h \to 0$, (4.10) can be expressed as $M = Lf$ [24], where f is the fraction of the loops that lies beneath the slots. Since in the considered geometry $f \approx 1$, in this hypothetic case, the magnetic flux linkage would be maximum, $M \approx L$.

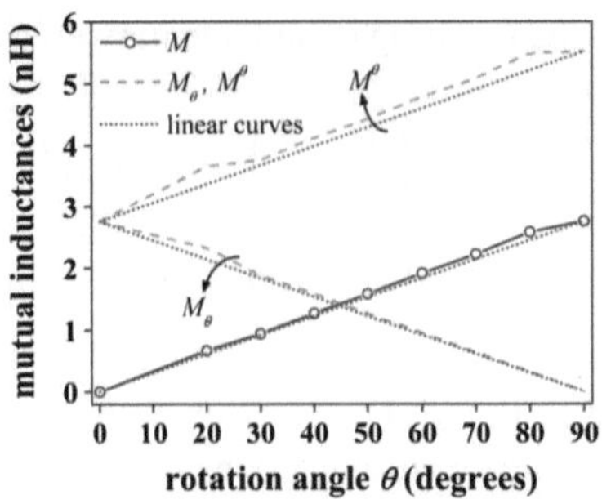

Fig. 4.44 Extracted mutual inductances of the models in Fig. 4.40 for the structure in Fig. 4.42 versus the rotation angle by steps of 10°. For $\theta = 0°$, it is assumed that $M = 0$ because the lack of a notch does not allow the circuit parameters to be extracted from the procedure in [10]. Similarly, for small angles ($\theta = 10°$), the parameters cannot be extracted either on account of the small M. The ideal linear curves are also plotted for comparison purposes. Reprinted with permission from [18]

mutual inductance can be interpreted as a differential mutual inductance, since it measures the difference between partial mutual inductances associated to each half of the structure. The areas and mutual inductances for the two extreme orientations are summarized in the caption of Fig. 4.41. Figure 4.44 shows the extracted values of M, M_θ, and M^θ for the structure in Fig. 4.42 for the full angular dynamic range. The most relevant result is that M varies continuously and quasi-linearly as a consequence of the circular shape of the structure, as expected. Hence, this phenomenology is indeed perfectly explained from (4.10).

While the notch magnitude of a lossless notch filter is infinite, in real circuits losses always prevent the attenuation from going to infinity (additionally, losses round the attenuation peak off, enhancing the rejection bandwidth around the notch frequency). The effect of losses on the notch that appears in the transmission coefficient is evaluated through a circuit model including losses, represented in Fig. 4.45a. It is assumed that power loss is dominated by resonator losses. Accordingly, losses are modeled to a good approximation by a series resistance R_s standing for the losses in the resonator or, equivalently by a parallel resistance R' in the transformed model in Fig. 4.45b [27, 28]. In the absence of losses, R_s is zero while R' is infinite. It is also important to emphasize that R' indicates that minimum losses in the equivalent circuit model are achieved by maximizing M and, obviously, by minimizing R_s (for a given resonance frequency, f_0). With respect to M, it is proportional to the area of the loops as expressed in (4.10), and it is also influenced by the distance between the coupled elements. Regarding R_s, in general, the larger the electrical size of the resonator, the higher the radiation losses but the smaller the conductor losses [32] (there is a tradeoff). Since commonly the radiation efficiency of a small loop antenna is small [29], the conductor loss of the ELC resonator is likely to be larger than the radiation loss. To sum up, what is to be expected is that the notch depth at a given

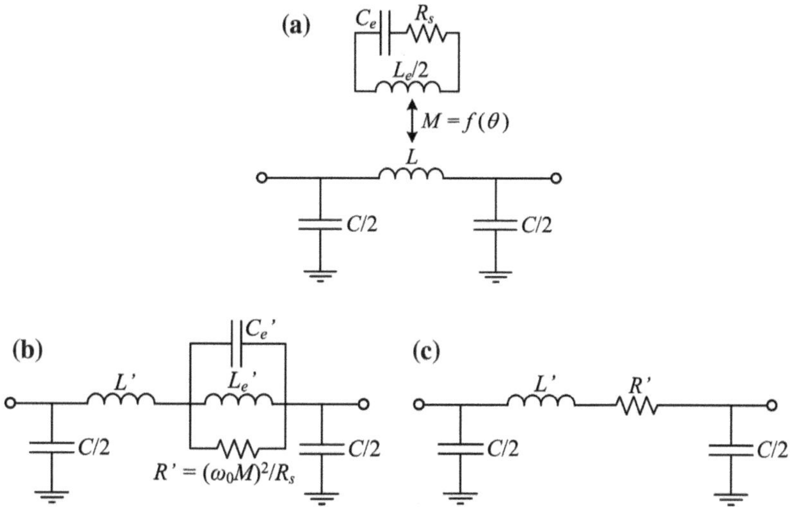

Fig. 4.45 **a** Equivalent circuit model of a CPW loaded with an ELC resonator, including resonator losses. **b** Transformed circuit model. **c** Transformed circuit model at the notch frequency

frequency increases with the electrical size of the resonator and with the proximity between the line and the resonator.

The resulting extracted resistances for the structure in Fig. 4.42 are shown in Fig. 4.46. As can be seen, R_s is roughly constant while R' grows strongly with the rotation angle due to its quadratic dependence on M, that also increases with the angle (Fig. 4.44). Figure 4.47 shows the transmission coefficient of the lossy electromagnetic and circuit simulations. As expected, the higher the parallel resistance, the deeper the notch magnitude.

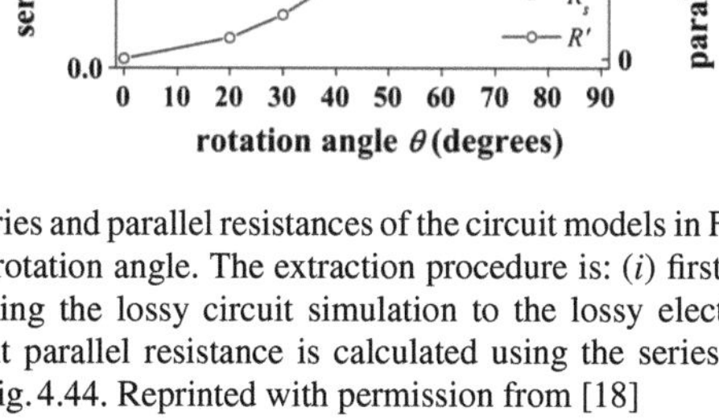

Fig. 4.46 Extracted series and parallel resistances of the circuit models in Fig. 4.45 for the structure in Fig. 4.42 versus the rotation angle. The extraction procedure is: (*i*) first, the series resistance is estimated by curve fitting the lossy circuit simulation to the lossy electromagnetic simulation; (*ii*) then, the equivalent parallel resistance is calculated using the series resistance, and the net mutual inductance of Fig. 4.44. Reprinted with permission from [18]

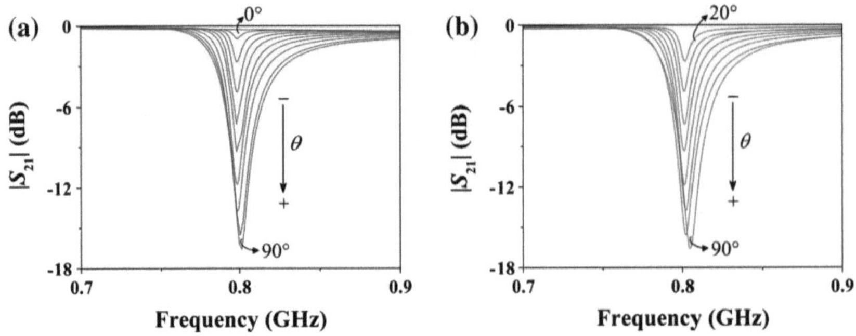

Fig. 4.47 Transmission coefficient obtained by **a** electromagnetic and **b** circuit simulations for the structure in Fig. 4.42. The circuit parameters for some of the orientations are given in Table 4.1 with $R_s \approx 0.49\,\Omega$. Reprinted with permission from [18]

The notch magnitudes observed in the transmission coefficients shown in Fig. 4.47 are plotted in Fig. 4.48. At the notch frequency, the resonator $L_e'C_e'$ opens and the resulting equivalent circuit is that of a conventional transmission line with R' in the series branch, as depicted in Fig. 4.45c. By the calculation of the transmission ($ABCD$) matrix of that circuit, the analytical expression for the magnitude of the transmission coefficient is found to be [1]

$$|S_{21}|\,(\text{dB}) = 20\log_{10}\left|\frac{1}{Z_s\left(\frac{Y_p^2 Z_0}{2}+\frac{1}{2Z_0}\right)+Y_p\left(Z_s+Z_0\right)+1}\right|, \tag{4.12}$$

where $Z_s = j\omega L' + R'$ is the impedance of the series branch, $Y_p = j\omega C/2$ is the admittance of the shunt branch, and Z_0 is the port impedance. According to (4.12), the line elements in the presence of the ELC resonator (L' and C) influence the notch depth. Nevertheless, for electrically small structures, the main signal rejection

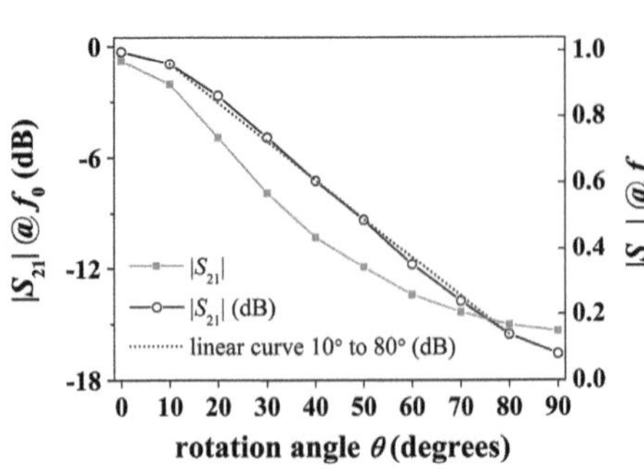

Fig. 4.48 Notch magnitude at the notch frequency, f_0, in the structure shown in Fig. 4.42 versus the rotation angle. Reprinted with permission from [18]

mechanism is due to R' rather than to these line elements. In the absence of these line elements ($Z_s = R'$ and $Y_p = 0$), the notch magnitude reduces to

$$|S_{21}|\,(\text{dB}) = 20\log_{10}\left|\frac{1}{1 + R'/2Z_0}\right|, \tag{4.13}$$

imposing a boundary dictating the minimum attenuation. Specifically, at very low frequencies, where the line elements do not play a significant role, (4.12) converges to (4.13). Instead of calculating the analytical sensitivity, in the light of the numerical solutions in Fig. 4.48, the magnitude of the notch in logarithmic scale is almost linearly dependent with the rotation angle, and such a property is interesting for sensing purposes (Sect. 6.5). To sum up, the proposed lossy equivalent circuit model explains that the notch magnitude depends not only on losses, but also on the line-to-resonator mutual inductance that, in turn, depends on the rotation angle of the resonator (the notch bandwidth broadens as the magnetic coupling strengthens, enhancing the notch depth). Theory and simulated results are in good agreement with experimental data, provided in Sect. 6.5.2.

4.3.4 MLC-Loaded Microstrip Lines

As compared to other slot resonators, apart from a magnetic wall, the MLC resonator exhibits an electric wall (Sect. 2.3.5). In [33] a microstrip line is loaded with an MLC resonator etched in the ground plane, assuming that the resonator symmetry plane exhibiting an electric wall is aligned with the line, as depicted in Fig. 4.49. If the (unusual) electric wall of the MLC is aligned with the symmetry plane of the microstrip line (a magnetic wall), there is an all-pass response. By contrast, when the MLC is displaced laterally, a notch arises, because the symmetry necessary to prevent from the resonance condition is not satisfied anymore.

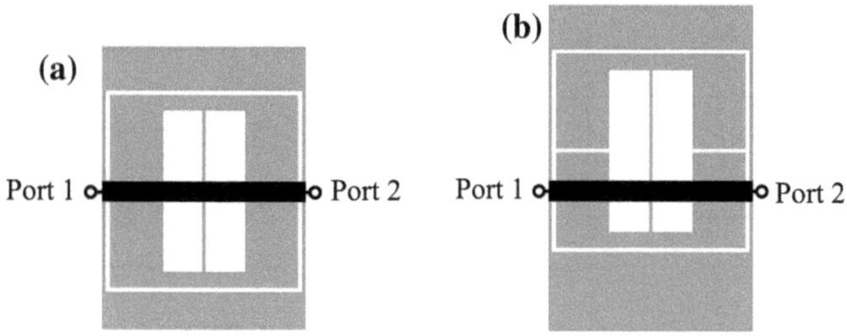

Fig. 4.49 Microstrip line loaded with an MLC with its electric wall **a** aligned and **b** laterally misaligned with the line. The substrate is *Rogers RO3010* with thickness $h = 1.27$ mm and relative permittivity $\epsilon_r = 11.2$. The line width is $W = 1$ mm (50-Ω line without the presence of the MLC). The MLC dimensions are (see Fig. 2.25): $w_1 = 8$ mm, $l_1 = 2$ mm, $w_2 = s = 0.2$ mm and $l_2 = l_3 = 10$ mm

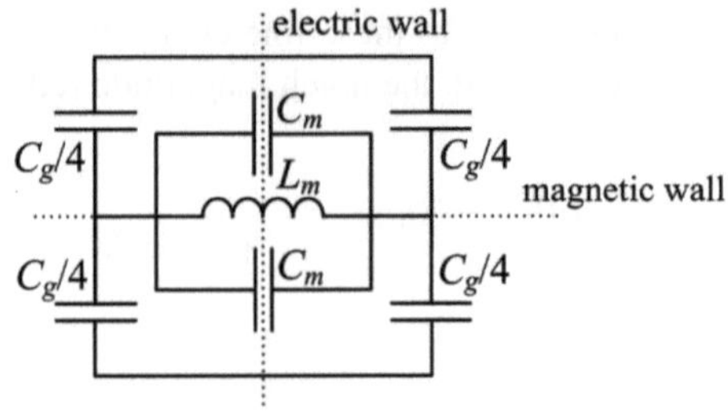

Fig. 4.50 Proposed equivalent circuit model of an MLC resonator. The inductance L_m accounts for the inductance through the path connecting the two inner metallic regions; this inductance is parallel-connected to the capacitances present across the slots between the two inner halves, denoted by C_m; finally, C_g stands for the edge capacitance of the external slot ring, and is split into four quadrants

An equivalent circuit model for an isolated MLC resonator is presented, and it is shown in Fig. 4.50. As it must be, the circuit model is perfectly bisymmetric. The proposed equivalent circuit model of the considered MLC-loaded line is represented in Fig. 4.51a, where the MLC is electrically coupled to the line as a function of its position (by C_1 and C_2). In the case where the resonator is aligned with the line, a transmission line model is obtained, as depicted in Fig. 4.51b. Contrarily, when the MLC and the line are no longer aligned, the MLC can be excited by the electric field generated by the line and a notch in the transmission coefficient occurs. The effect of this symmetry disruption produced by shifting laterally the MLC can be taken into account in the model by considering the coupling capacitances (i.e. C_1 and C_2) as variable capacitances (more precisely, the line inductance L is also influenced by the resonator position, though it is not explicitly considered tunable). From the circuit model viewpoint, these asymmetric values of the coupling capacitances allow

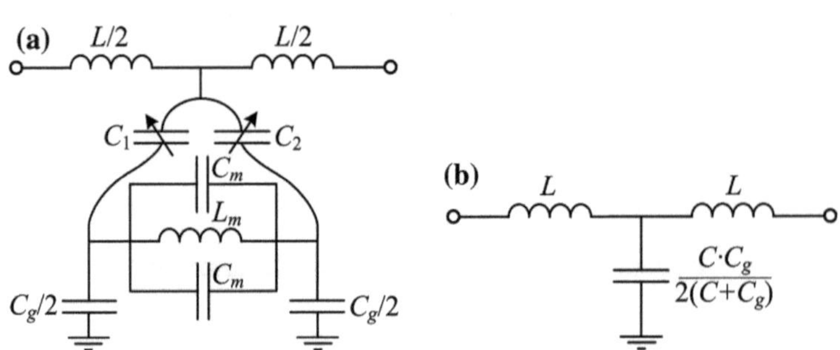

Fig. 4.51 **a** Equivalent circuit model of a microstrip line loaded with an MLC whose electric wall is aligned or laterally displaced with the line. **b** Circuit model by applying the magnetic wall concept under the assumption of alignment ($C_1 = C_2 = C/2$). The inductance L stands for the line inductance, $C = C_1 + C_2$ models both the line capacitance and the coupling capacitance with the resonator, where C_i ($i = 1, 2$) are the capacitances on each side in regard to the line axis; the resonator is modeled by the inductance L_m, and the capacitances C_m and C_g

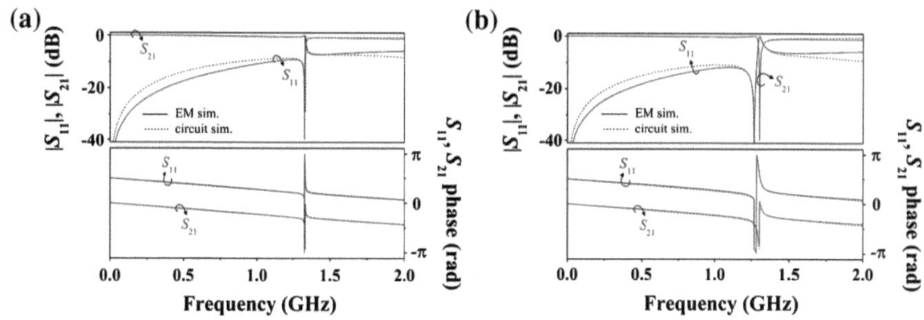

Fig. 4.52 Frequency response for the structure that results when the MLC in Fig. 4.49a is laterally shifted **a** 0.5 mm and **b** 2 mm, and for its circuit model in Fig. 4.51a. The baseline circuit parameters are those given in the caption of Fig. 4.22 corresponding to the circuit in Fig. 4.21a, since the same dimensions and substrate are considered. Afterward, the variable coupling capacitances, which depend on the amount of displacement, are adjusted by curve fitting to the electromagnetic simulation. The other elements are nearly unaltered (a slight optimization may be required), with the exception of the line inductance which is also perturbed by the position of the slotted resonator. The circuit parameters are: $L_m = 3.71$ nH, $C_g = 9.5$ pF; for 0.5 mm, $L = 9.48$ nH, $C_1 = 0.89$, $C_2 = 0.36$ ($C = 1.25$), and $C_m = 0.59$ pF; for 2 mm, $L = 8.94$ nH, $C_1 = 1.36$, $C_2 = 0.083$ ($C = 1.44$ pF) and $C_m = 0.66$ pF. Reprinted with permission from [33]

current flowing through the MLC inductance and capacitance, and hence the particle is electrically driven by the line.

Note that as long as the symmetry plane of the MLC is sufficiently apart from the symmetry plane of the line, the reported equivalent circuit model coincides with that of MLC-loaded differential microstrip lines shown in Fig. 4.21a in the absence of one of the microstrip lines of the differential line. It is also worth mentioning that the circuit model is similar to that of FSIR-loaded CPWs when the magnetic coupling is neglected (see Fig. 4.33 in Sect. 4.3.1).

The circuit model is validated for two different displacements, as shown in Fig. 4.52. It is seen that the lateral shift of the MLC can be basically taken into account by modifying the coupling capacitances and the line inductance. The higher the displacement, the stronger the notch, as is illustrated in Fig. 4.52. The measurements of a frequency-coded barcode (Sect. 7.4.2) are also consistent with the theoretical arguments provided here.

4.4 On the Generalization of Symmetry Rupture

This section aims at the generalization of the symmetry properties in transmission lines loaded with resonant particles, regardless of the number of (per-unit-cell) resonators and the symmetry disruption mechanism.

In the previous section, symmetry rupture in a transmission line loaded with a symmetric resonator was forced by misaligning the symmetry planes of the two

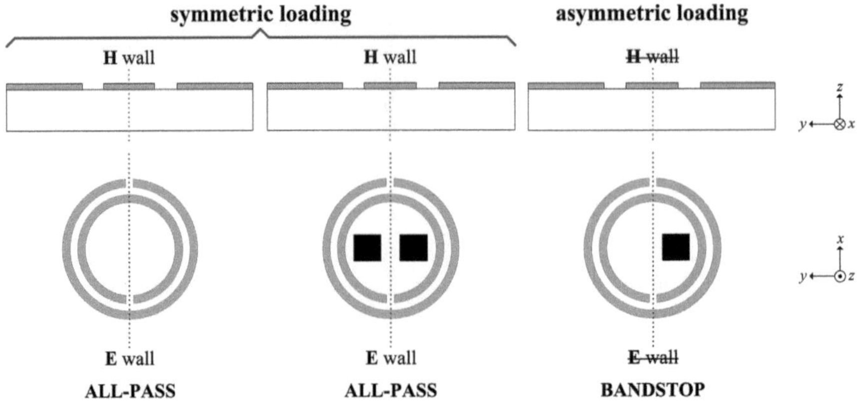

Fig. 4.53 Illustration of symmetry rupture by square inclusions in the approach based on single resonators, where a CPW is loaded with an SRR

elements. Accordingly, the structure is symmetric or asymmetric depending on whether the resonator is aligned or not with the line, respectively. However, the lack of symmetry can be caused by mechanisms different from the relative orientation between the line and the resonator. For instance: (*i*) the resonator symmetry may be topologically destroyed; (*ii*) other elements can be included (e.g. dielectric loads); or (*iii*) the surrounding medium can be subjected to inhomogeneities. In those cases where symmetry rupture is not based on geometrical misalignments, the propagating characteristics of the structure should be defined in a general sense in terms of symmetry/asymmetry rather than on alignment/misalignment. *Provided that a transmission line is loaded with a resonator, and that the line may also be loaded with other inclusions* (see Fig. 4.53).

- *If the whole structure is symmetric (for which the line and the resonator are necessarily symmetric, and whose symmetry planes must coincide), if the symmetry planes of the line and resonator are of:*
  - *Distinct nature, resonance is inhibited because the net line-to-resonator coupling is null* ($|S_{21}| \approx 1$).
  - *The same nature, resonance is established because the resonator is tightly coupled to the line* ($|S_{21}| \approx 0$).
- *If the whole structure is asymmetric, resonance is allowed, the stopband characteristics being dependent on the asymmetry that determines the line-to-resonator coupling* ($0 < |S_{21}| < 1$).

Hence, the stopband characteristics may be modulated by symmetry properties as a whole, that determine the net coupling between the line and the resonator. It is for this reason that, in this thesis, the principle of operation in transmission lines loaded with single symmetric resonators is said to be based on a *coupling-modulated*

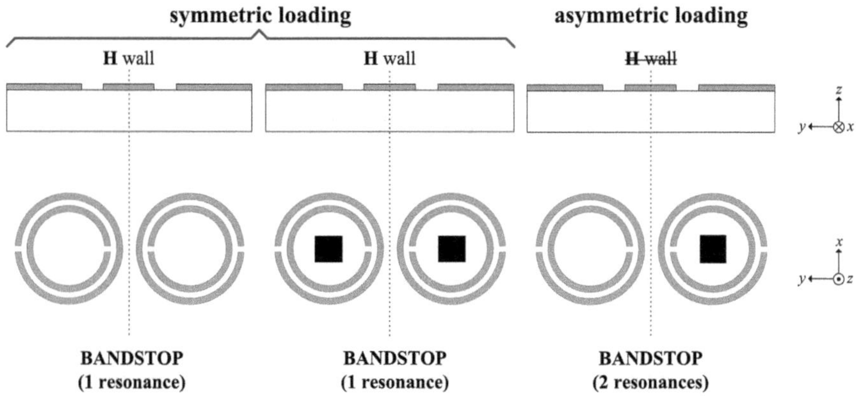

Fig. 4.54 Illustration of symmetry rupture by square inclusions in the approach based on pairs of resonators, where a CPW is loaded with a pair of SRRs. Note that the symmetry planes of the SRRs are not required to be parallel to the longitudinal x-direction; if the slits of the SRRs are oriented in perpendicular to the longitudinal axis, CPW-to-SRR mixed coupling is prevented

resonance. For the particular case of resonator-loaded lines with bimodal even/odd-mode propagation, the functionality termed *selective mode suppression* is derived; the coupling is null for one mode, whereas it is tight for the other.

Thus far, the symmetry properties of loaded transmission lines have been analyzed considering a single (per-unit-cell) resonator. In addition, symmetry properties in transmission lines loaded with any other number of resonators (arranged transversely to the line) can also be inferred. This book considers the particular case with pairs of resonators.

The generalization of symmetry rupture for pairs of resonators is subjected to using identical resonators (symmetric resonators are not necessary) in order to possess a plane of symmetry (midway between the resonators).[18] Thereby the loaded line can be symmetric by aligning the plane of symmetry of the paired resonators with the line. In terms of absolute symmetry, *provided that a transmission line is loaded with a pair of resonators, and that the line may also be loaded with other inclusions* (*see Fig.* 4.54).

- *If the whole structure is symmetric (where the line must be symmetric and the pair of resonators must be identical and symmetrically oriented with regard to the line axis) there is one transmission zero because the resonance frequencies of the two resonators degenerate to the same one* ($|S_{21}| \approx 0$ @ f_0).
- *If the whole structure is asymmetric, there are in general*[19] *two transmission zeros because two different resonances arises* ($|S_{21}| \approx 0$ @ f_1, f_2).

---

[18]This is the case, for instance, of classical CPWs loaded with pairs of SRRs.

[19]An asymmetric structure based on pairs of resonators may exhibit a single transmission zero, such as when two different uncoupled resonators have the same resonance frequency or when one of the resonances is prevented by aligning a symmetric resonator with the line.

Consequently, in this approach based on pairs of resonators, a high controllability in regard to the number and frequency of resonances as a function of symmetry may be achieved.[20] For this reason, in contrast to transmission lines loaded with single resonators, the working principle in transmission lines loaded with multiple resonators is based on *resonance frequency splitting/shifting*.[21]

In the approach based on multiple resonators, there are two fundamental differences as compared to using a single resonator. One of the differences concerns the fact that, if symmetric resonators are used, the two resonators cannot be simultaneously aligned with the line (we can align only one resonator). Therefore, only one resonance frequency can be inhibited at most in transmission lines loaded with pairs of resonators (and in the general case with more than a single resonator). The relevant conclusion is that resonance always occurs. The other interesting aspect of this approach is that there is no restriction in the orientation of the individual resonators in regard to the line. This is useful to cancel mixed-coupling effects in certain topologies. For instance, as is illustrated in Fig. 4.54, if symmetric SRRs are considered, their symmetry planes are not needed to be oriented longitudinally and the SRRs can be coupled only magnetically to the line (provided that the slits are oriented in the transverse direction).

Symmetry rupture by adding inclusions to the structure may be useful for the implementation of differential sensors and comparators. With single resonators, symmetry has been broken only by misalignments so far (Sect. 4.3). Using pairs of resonators, symmetry disruption is explored by changing the topology of the resonators or by introducing external inclusions. The corresponding results for the considered implementations are summed up in the next subsections. To the author's best knowledge, the first reported structure where symmetry is broken by means of different dimensions of a pair of resonators was used as a sensing device for displacements by Horestani et al. [34]. On the other hand, the first transmission line loaded with pairs of resonators whose symmetry is perturbed by means of inclusions was proposed as a differential sensor and comparator for dielectric measurements by Naqui et al. in [35] (Sects. 4.4.1 and 6.7.1). In both sensors, sensing is based on the split/shift of resonance frequencies.

4.4.1 Microstrip Lines Loaded with Pairs of SISSs

This section is devoted to microstrip lines loaded with pairs of SISSs (stepped-impedance shunt-stubs), previously introduced in Sect. 3.3.1. Figure 4.55 shows a

[20] As a secondary effect, the bandwidth and depth of the notches may also depend on symmetry.

[21] Resonance frequency shifting makes reference to the fact that resonance frequencies are shifted, regardless of what produces the shifting. On the other hand, resonance frequency splitting means that the intrinsic resonance frequency of identical resonators is split into different frequencies in the case that identical resonators are coupled. A clear example is the frequency splitting in transmission lines periodically loaded with coupled resonators (Sect. 3.2); there is frequency shifting and, more precisely, frequency splitting.

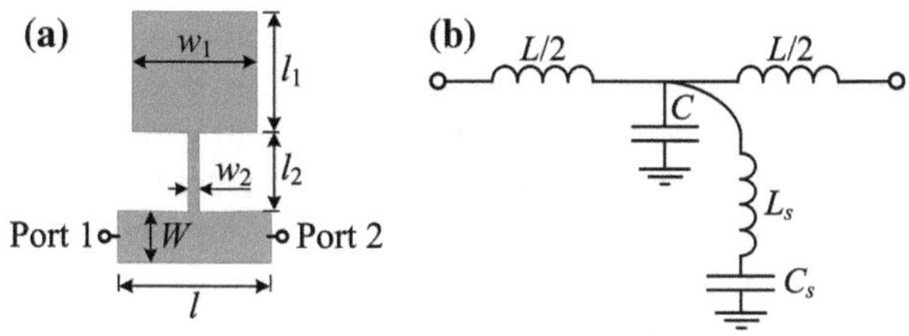

Fig. 4.55 **a** SISS-loaded microstrip line and **b** equivalent circuit model. The inductance and capacitance of the line section are L and C, respectively, whereas the SISS inductance and capacitance, L_s and C_s, are primarily determined by the dimensions of the narrow and wide strips, respectively (Sect. 3.3.1)

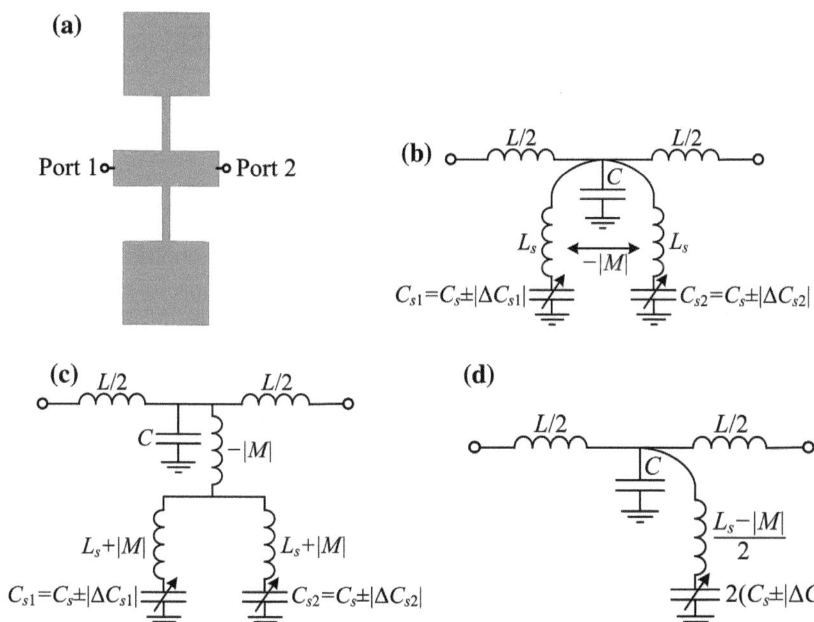

Fig. 4.56 **a** Microstrip line loaded with a pair of symmetrically loaded SISSs. **b** Equivalent circuit model where capacitive perturbations are applied to the SISSs. **c** Transformed equivalent circuit model. **d** Transformed equivalent circuit model in the case of symmetric perturbations

microstrip line loaded with a single SISS and its equivalent circuit model. As compared to SIRs or FSIRs (Sect. 4.3.2), SISSs can be loaded to a conventional microstrip line and are excited by port terminals (directly connected to the line) rather than by electromagnetic coupling.[22]

Naqui et al. investigate in [35] the behavior of a microstrip line loaded with a pair of mirrored SISSs, as depicted in Fig. 4.56a. Because the two SISSs are loaded

---

[22]Consequently, the resonance frequency cannot be prevented in transmission lines loaded with single SISSs.

to the same junction, the resonators are close to each other, and they are seen to be magnetically coupled. Since the resonators are loaded on opposite sides with respect to the microstrip line, the currents in their junction are anti-parallel, and the magnetic coupling is negative. The derived equivalent circuit model is shown in Fig. 4.56b, which can be transformed using an equivalent T-circuit (Appendix B) to that depicted in Fig. 4.56c. For the purpose of designing permittivity sensors (Sect. 6.7.1), each individual resonator is capacitively-perturbed by ΔC_{si} ($i = 1, 2$). Therefore, two simultaneous capacitive perturbations are included. When the structure is symmetric (identical perturbations, $\Delta C_{s1} = \Delta C_{s2} = \Delta C_s$), the circuit depicted in Fig. 4.56d results. By symmetry, there is a single transmission zero at an angular resonance frequency[23]

$$\omega_e = \frac{1}{\sqrt{(L_s - |M|)(C_s \pm |\Delta C_s|)}} = \frac{\omega_0}{\sqrt{1 - |k_m|}}, \tag{4.14}$$

where $k_m = -|M|/L_s$ is the magnetic coupling coefficient [32] and

$$\omega_0 = \frac{1}{\sqrt{L_s(C_s \pm |\Delta C_s|)}}, \tag{4.15}$$

is the resonance frequency in the absence of inter-resonator coupling, that is, the resonance frequency of a microstrip line loaded with a single capacitively-perturbed SISS. Note that the negative inter-resonator coupling shifts the resonance frequency upward. On the contrary, when symmetry is disrupted (different capacitive perturbations), two transmission zeros appear. Interestingly, due to the inter-resonator coupling, resonance frequency splitting[24] occurs [32]. By forcing the shunt branch impedance to be zero, the two resonance frequencies in the asymmetric case are [32]

$$\omega_{1,2} = \sqrt{\frac{L_s(C_{s1} + C_{s2}) \pm \sqrt{[L_s(C_{s1} - C_{s2})]^2 + 4C_{s1}C_{s2}M^2}}{2C_{s1}C_{s2}\left(L_s^2 - M^2\right)}}, \tag{4.16}$$

that depend on the capacitive asymmetry as well as on the resonator-to-resonator coupling.

In order to validate the circuit models, symmetry is broken by changing the length of one of the wide sections of the SISS pair. Three microstrip-based structures are considered: one with a symmetric pair of SISSs, another with an asymmetric pair by increasing l_1 in one of the SISSs ($+|\Delta C_{s2}|$), and another one with an asymmetric

---

[23] This resonance frequency is denoted by ω_e following the nomenclature in [32] to emphasize the presence of an open circuit at the symmetry plane of two identical coupled resonators produced by an even mode excitation.

[24] In contrast to the resonance frequency splitting in Sect. 3.2 where the coupled resonators are identical, in the present situation two different resonant structures (identical resonators with different perturbations) are necessary for frequency splitting. Therefore, in microstrip lines loaded with pairs of SISS, the resonance frequency splitting actually shifts the two (already different) resonance frequencies enhancing the separation between them.

Fig. 4.57 Photograph of the fabricated microstrip lines loaded with pairs of SISSs. The dimensions are: $W = 1.83\,\text{mm}$ (50-Ω line), $l = 15.9$, $l_1 = l_2 = 2.6$, $\Delta l_1 = \pm 0.5$, $w_1 = 5.5\,\text{mm}$, and $w_2 = 250\,\mu\text{m}$. The substrate is *Rogers RO4003C* with thickness $h = 812.8\,\mu\text{m}$, $\epsilon_r = 3.38$ (the considered relative permittivity is $\epsilon_r = 3.1$ to fit better the electromagnetic simulation with the measurements), and loss tangent $\tan\delta = 0.0021$

pair by decreasing l_1 ($-|\Delta C_{s2}|$). These structures are fabricated and measured, as shown in Figs. 4.57 and 4.58a, respectively. As can be noticed, asymmetry leads to the appearance of a lower resonance frequency, whose attenuation is lower than that associated to the upper one. The circuit model verification is depicted in Fig. 4.58b, where good agreement with measurements can be appreciated. Further analysis of microstrip lines loaded with pairs of SISSs toward the implementation of sensors is provided in Sect. 6.7.1.

4.4.2 Coplanar Waveguides Loaded with Pairs of SRRs

It is worth mentioning that Naqui et al. also analyzed CPWs loaded with pairs of SRRs assuming both equal and different resonators [36, 37]. This analysis is part of the latest research on transmission lines loaded with pairs of resonators, and it is for this reason that the results are not included in this book. Instead, a summary is reported here.

In the above-cited references, in order to avoid mixed coupling in the line-to-resonator interaction, the slits of the SRRs are oriented along the transverse direction. Furthermore, the slits of the larger ring forming the SRRs are located on opposite sides to ensure that the inter-resonator coupling is dominantly magnetic (obviously, as the distance between the SRRs decreases, the inter-resonator magnetic coupling strengthens). Indeed, inter-resonator coupling is generally expected in CPWs loaded with pairs of SRRs, despite the fact that it has not been modeled in the literature to date (to the author's knowledge). An equivalent circuit model is presented, whose transmission zero frequencies are found to be given by the same expressions as those obtained in microstrip lines loaded with pairs of SISSs. In these two quite different structures, the phenomenology associated to the effects of inter-resonator coupling is almost the same. One exception is the sign of the magnetic coupling (negative in microstrip lines loaded with pairs of SISSs and positive in CPW loaded with pairs of SRRs). Another difference is the case of distinct resonators but exhibiting the same

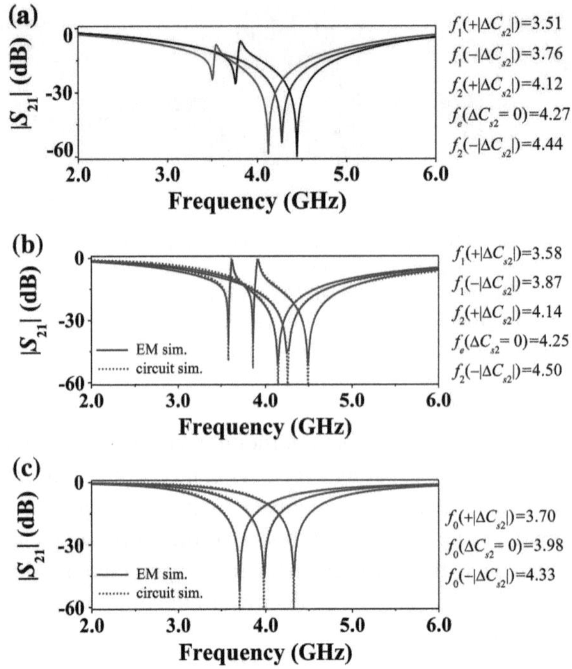

Fig. 4.58 Transmission coefficient magnitude in the microstrip lines loaded with pairs of SISSs under study. **a** Measured transmission coefficients. **b** Transmission coefficients of the lossless electromagnetic and circuit simulations. **c** Transmission coefficients of the lossless electromagnetic and circuit simulations using a single SISS (not fabricated, but used in the extraction procedure of the circuit elements). The lumped element values are extracted as follows. (*i*) The parameters in Fig. 4.55b are extracted loading a single SISS without any perturbation using the procedure in [9] adapted to the considered circuit; $L = 1.80\,\text{nH}$, $C = 0.57\,\text{pF}$, $L_s = 2.46\,\text{nH}$ and $C_s = 0.65\,\text{pF}$. (*ii*) Each perturbation is extracted using (4.15) from the electromagnetic simulation employing a single perturbed SISS; $\Delta C_{s2} = \pm 0.15 C_s = \pm 0.10\,\text{pF}$. (*iii*) $|M| = 0.31\,\text{nH}$ ($k_m = -|M|/L_s = -0.13$) is obtained by curve fitting the circuit simulation of the model in Fig. 4.56b to the electromagnetic simulation of the line loaded with symmetric SISSs; alternatively, (4.14) or (4.16) can also be used for a symmetric or an asymmetric pair, respectively. Reprinted with permission from [35]

resonance frequency. In both structures, two resonance frequencies appear. However, for SRR-loaded lines there is an exceptional balance condition that nulls one of these frequencies, this being not possible in SISS-loaded lines.

References

1. D.M. Pozar, *Microwave Engineering*, 3rd edn. (Wiley, New York, 2005)
2. T.C. Edwards, M.B. Steer, *Foundations of Interconnect and Microstrip Design* (Wiley, New York, 2000)

3. K.C. Gupta, R. Garg, I.J. Bahl, P. Bhartia, *Microstrip Lines and Slotlines* (Artech House, Boston, 1996)
4. R.K. Mongia, J. Hong, P. Bhartia, I.J. Bahl, *RF and Microwave Coupled-Line Circuits* (Artech House, Boston, 1999)
5. R.N. Simons, *Coplanar Waveguide Circuits, Components, and Systems* (Wiley, New York, 2001)
6. F. Martín, *Artificial Transmission Lines for RF and Microwave* (Wiley, New York, 2015)
7. W.-T. Liu, C.-H. Tsai, T.-W. Han, T.-L. Wu, An embedded common-mode suppression filter for GHz differential signals using periodic defected ground plane. IEEE Microw. Wirel. Compon. Lett. **18**(4), 248–250 (2008)
8. J. Naqui, M. Durán-Sindreu, F. Martín, Selective mode suppression in coplanar waveguides using metamaterial resonators. Appl. Phys. A **109**(4), 1053–1058 (2012)
9. J. Bonache, M. Gil, I. Gil, J. García-García, F. Martín, On the electrical characteristics of complementary metamaterial resonators. IEEE Microw. Wirel. Compon. Lett. **16**(10), 543–545 (2006)
10. F. Aznar, M. Gil, J. Bonache, L. Jelinek, J.D. Baena, R. Marqués, F. Martín, Characterization of miniaturized metamaterial resonators coupled to planar transmission lines through parameter extraction. J. Appl. Phys. **104**, 114501 (2008)
11. J. Naqui, M. Durán-Sindreu, F. Martín, Modeling split-ring resonator (SRR) and complementary split-ring resonator (CSRR) loaded transmission lines exhibiting cross-polarization effects. IEEE Antennas Wirel. Propag. Lett. **12**, 178–181 (2013)
12. J. Naqui, M. Durán-Sindreu, A. Fernández-Prieto, F. Mesa, F. Medina, F. Martín, Differential transmission lines loaded with split ring resonators (SRRs) and complementary split ring resonators (CSRRs), in *Metamaterials '2011: the 5th International International Congress on Advanced Electromagnetic Materials in Microwaves and Optics*, Barcelona, October 2011
13. J. Naqui, A. Fernández-Prieto, M. Durán-Sindreu, J. Selga, F. Medina, F. Mesa, F. Martín, Split rings-based differential transmission lines with common-mode suppression, in *IEEE MTT-S International Microwave Symposium Digest*, Baltimore, June 2011
14. J. Naqui, A. Fernández-Prieto, M. Durán-Sindreu, F. Mesa, J. Martel, F. Medina, F. Martín, Common-mode suppression in microstrip differential lines by means of complementary split ring resonators: Theory and applications. IEEE Trans. Microw. Theory Tech. **60**(10), 3023–3034 (2012)
15. J. Naqui, M. Durán-Sindreu, F. Martín, Selective mode suppression in microstrip differential lines by means of electric-LC (ELC) and magnetic-LC (MLC) resonators, in *4th International Conference on Metamaterials, Photonic Crystals and Plasmonics (META)*, Sharjah, March 2013
16. J. Naqui, M. Durán-Sindreu, F. Martín, Selective mode suppression in microstrip differential lines by means of electric-LC (ELC) and magnetic-LC (MLC) resonators. Appl. Phys. A **115**, 637–643 (2014)
17. D. Schurig, J. Mock, D. Smith, Electric-field-coupled resonators for negative permittivity metamaterials. Appl. Phys. Lett. **88**(041109) (2006)
18. J. Naqui, F. Martín, Transmission lines loaded with bisymmetric resonators and their application to angular displacement and velocity sensors. IEEE Trans. Microw. Theory Tech. **61**(12), 4700–4713 (2013)
19. J. Naqui, M. Durán-Sindreu, F. Martín, Novel sensors based on the symmetry properties of split ring resonators (SRRs). Sensors—Special Issue Metamaterials for Sensing **11**(8), 7545–7553 (2011)
20. J. Naqui, M. Durán-Sindreu, F. Martín, Alignment and position sensors based on split ring resonators. Sensors—Special Issue Ultra-Small Sensor Systems and Components **12**(9), 11 790–11 797 (2012)
21. J. Naqui, M. Durán-Sindreu, F. Martín, On the symmetry properties of coplanar waveguides loaded with symmetric resonators: analysis and potential applications, in *IEEE MTT-S International Microwave Symposium Digest*, Montreal, June 2012

22. J. Naqui, F. Martín, Mechanically reconfigurable microstrip lines loaded with stepped impedance resonators and potential applications, Int. J. Antennas Propag. **2014**(346838) (2014)
23. J. Naqui, M. Durán-Sindreu, F. Martín, Transmission lines loaded with bisymmetric resonators and applications, in *IEEE MTT-S International Microwave Symposium Digest*, Seattle, June 2013
24. F. Martín, J. Bonache, F. Falcone, M. Sorolla, R. Marqués, Split ring resonator-based left-handed coplanar waveguide. Appl. Phys. Lett. **83**(22), 4652–4654 (2003)
25. F. Martín, F. Falcone, J. Bonache, R. Marqués, M. Sorolla, Miniaturized coplanar waveguide stop band filters based on multiple tuned split ring resonators. IEEE Microw. Wirel. Compon. Lett. **13**(12), 511–513 (2003)
26. F. Aznar, J. Bonache, F. Martín, Improved circuit model for left-handed lines loaded with split ring resonators, Appl. Phys. Lett. **92**(043512) (2008)
27. R. Marqués, F. Martín, M. Sorolla, *Metamaterials with Negative Parameters: Theory, Design and Microwave Applications*. (Wiley, New York, 2008)
28. I.J. Bahl, P. Bhartia, *Microwave Solid State Circuit Design* (Wiley, New York, 2003)
29. C.A. Balanis, *Antenna Theory: Analysis and Design*, 2nd ed. (Wiley, New York, 1997)
30. R. Marqués, F. Medina, R. Rafii-El-Idrissi, Role of bianisotropy in negative permeability and left-handed metamaterials, Phys. Rev. B **65**(144440) (2002)
31. J.R. Reitz, F.J. Milford, R.W. Christy, *Foundations of Electromagnetic Theory*, 4th edn. (Addison-Wesley Publishing Company, New York 1996)
32. J.S. Hong, M.J. Lancaster, *Microstrip Filters for RF/Microwave Applications*. (Wiley, New York, 2001)
33. J. Naqui, M. Durán-Sindreu, F. Martín, Differential and single-ended microstrip lines loaded with slotted magnetic-LC (MLC) resonators, Int. J. Antennas Propag. **2013**(640514) (2013)
34. A.K. Horestani, J. Naqui, Z. Shaterian, D. Abbott, C. Fumeaux, F. Martín, Two-dimensional alignment and displacement sensor based on movable broadside-coupled split ring resonators. Sens. Actuat. A Phys. **210**, 18–24 (2014)
35. J. Naqui, C. Damm, A. Wiens, R. Jakoby, L. Su, F. Martín, Transmission lines loaded with pairs of magnetically coupled stepped impedance resonators (SIRs): Modeling and application to microwave sensors, in *IEEE MTT-S International Microwave Symposium Digest*, Seattle June 2014
36. J. Naqui, L. Su, J. Mata, F. Martín, Analysis of transmission lines loaded with pairs of coupled resonant elements and application to sensors, in *Moscow International Symposium on. Magnetism (MISM-2014)*, Moscow, June 2014
37. J. Naqui, L. Su, J. Mata, F. Martín, Analysis of transmission lines loaded with pairs of coupled resonant elements and application to sensors. J. Magn. Magn. Mater. **383**(4), 144–151 (2015)

Chapter 5
Application of Symmetry Properties to Common-Mode Suppressed Differential Transmission Lines

The selective mode suppression inspired by the alignment of symmetry planes described in Sect. 4.2 can be useful for the design of microwave differential circuits and components. Specifically, differential structures with common-mode suppression can be implemented. In the present chapter, we focus on common-mode suppressed differential lines in microstrip technology. As is typically required, the proposed structures suppress the common-mode propagation efficiently by means of the loading resonators, whereas differential-mode signals are nearly unaffected by the presence of the resonators. The key aspect to suppressing the common mode whilst keeping the differential mode almost unaltered is the perfect alignment between the line and the resonators.

Section 5.1 introduces the state-of-the-art of common-mode suppressed filters. Section 5.2 reviews briefly the principle of selective even/odd mode filtering. Afterward, in Sect. 5.3 we report an approach and some prototypes of common-mode suppressed differential microstrip lines.

5.1 Introduction

Differential or balanced[1] circuits have been important in electronic and communication systems for more than 50 years. Historically, they have been used in low frequency applications rather than in radio frequency (RF) and microwave systems. However, in last years, the demand for differential RF/microwave circuits has increased at a huge rate. The reason for such a trend is that differential signals offer high noise immunity and low electromagnetic interference (EMI) [2, 3], as compared to single-ended signals (Appendix D). These characteristics are excellent benefits for

[1] *Balanced* is the microwave term for *differential* [1].

J. Naqui, *Symmetry Properties in Transmission Lines Loaded with Electrically Small Resonators*, Springer Theses,
DOI 10.1007/978-3-319-24566-9_5

modern high-speed digital circuits and communication transceivers. Nonetheless, since most practical implementations of balanced lines incorporate a ground plane (e.g. a symmetric pair of microstrip lines), balanced systems commonly face the problematic common-mode signals.[2] Since in differential circuits the noise pickup comes almost entirely from common-mode signals, these signals are also termed *common-mode noise*. As is well known, common-mode signals may degrade the differential signal quality and produce unwanted radiated emissions as well [2].

As was introduced in Sect. 4.1, for the characterization of differential circuits, the definition of the differential- and common-mode is more convenient than that of the odd- and even-mode. For this reason, in microwave engineering, differential structures are usually characterized by mixed-mode scattering parameters (S-parameters), rather than by standard single-ended S-parameters. Thereby the transmission and reflection coefficients for the differential and common modes can be obtained. Nevertheless, there is a simple linear relationship between the mixed-mode and the standard S-parameters (Appendix D).

The common-mode choke is one of the most typical methods to suppress the common mode. Unfortunately, the choke works properly only at the MHz frequency range, and it does not fit today's small area requirement [4]. In this context, the design of differential lines in the GHz range able to suppress the common-mode noise while keeping the integrity of differential signals is of great importance. For such filters, there are several specifications, typically high rejection level and broadband suppression for the common-mode noise, and low insertion loss and phase distortion for differential signals [5].[3] For GHz differential signals, common-mode filters based on negative permittivity structures [4] or multilayer LTCC (low temperature co-fired ceramic) [6] have been reported. These structures are compact and provide efficient common-mode rejection over wide frequency bands, but are technologically complex. There have also been presented a number of approaches for the design of differential lines with common-mode suppression based on defected ground structures (DGSs); dumbbell-shaped periodic patterns etched symmetrically in the ground plane, underneath the differential lines, have been used to suppress the common mode [7]; using U-shaped and H-shaped coupled resonators etched symmetrically in the ground plane have also been used to achieve a wide stopband for the common mode [8]. Further structures have been reported in the last yew years [9–12], and a comprehensive review of strategies for common-mode suppression has been published [13].

Before ending this section, it is important to point out that, apart from common-mode suppressed differential lines, other balanced circuits (such as bandpass filters or power dividers) with common-mode rejection have also become a hot research topic recently [5, 10, 14–22].

[2]In the absence of a common ground plane, the differential transmission line becomes a two-conductor propagative structure where common-mode signals cannot be supported. For instance, coplanar strips (CPS) support only differential-mode signals.

[3]The common-mode rejection ratio (CMRR) may also be used as a figure of merit [5].

5.2 Symmetry-Based Selective Mode Suppression

Bimodal symmetric transmission lines, such as parallel microstrip lines and coplanar waveguides (CPWs) support even- and odd-mode propagation (Sect. 4.1). As was demonstrated in Sect. 4.2, we can suppress selectively either the even mode or the odd mode while keeping the other mode almost unaltered by loading symmetrically these lines with a resonator that exhibits a magnetic wall or an electric wall at resonance, respectively.

Unlike the broadband selective suppression of the differential mode (odd mode), which can be readily achieved by short-circuiting the differential signal,[4] the selective suppression in a wide spectral band of the common mode (even mode) is not so straightforward. Although symmetric CPWs can be used as propagative structures of odd-mode signals, they are rarely used for that purpose. Instead, symmetric parallel microstrip lines are preferred. Therefore, the focus here is not on suppressing the even mode in CPWs, but in differential microstrip lines where the odd mode is responsible for the propagation of the differential signaling. With a view to suppressing the even mode while keeping the integrity of the odd mode, the differential transmission line must be aligned with a resonator that exhibits a magnetic wall at its symmetry plane.

As far as the author knows, the selective mode suppression by aligning transmission lines with resonant elements was first proposed by Liu et al. [7] also for common-mode suppression in differential microstrip lines. Next sections are devoted to the considered resonators for achieving common-mode suppression, while maintaining the differential mode unaffected.

5.3 Common-Mode Suppressed Differential Microstrip Lines

5.3.1 CSRR- and DS-CSRR-Loaded Differential Microstrip Lines

Although a microstrip line is an unbalanced structure, a two-port balanced microstrip line can be designed by means of a symmetric pair of microstrip lines differentially excited.[5] A differential microstrip transmission line with common-mode suppression based on complementary split-ring resonators (CSRRs) is proposed and experimentally validated for the first time in [24], and further research is reported subsequently

[4]In CPW-based structures with asymmetric topologies, the rejection of the parasitic differential mode while remaining the common mode unaltered is of general interest. As was mentioned in Sect. 4.1, slot-mode rejection is usually and efficiently achieved using air bridges to short-circuit this mode [23].

[5]A balanced line carries balanced signals, that is, the two conductors carry signals with the same amplitude but with 180° phase shift; the signal on one conductor is referenced to the other one.

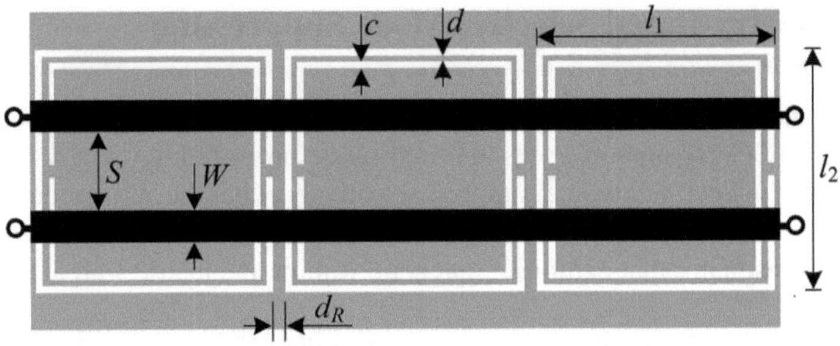

Fig. 5.1 Typical topology of a differential microstrip line loaded with square and coupled CSRRs symmetrically etched in the ground plane for common-mode suppression. The structure can be characterized in terms of four single-ended ports or two mixed-mode ports

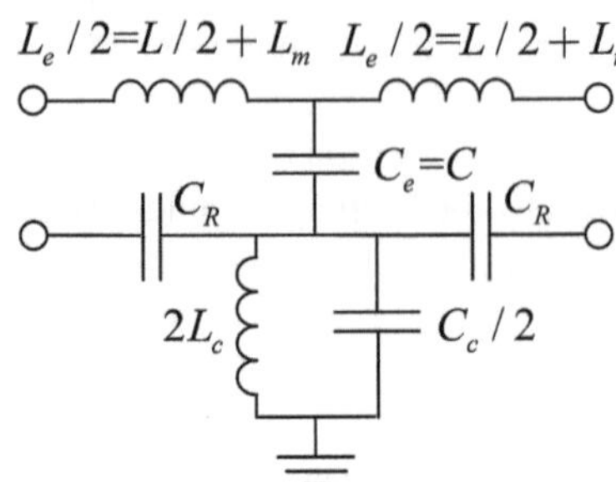

Fig. 5.2 Even-mode circuit model (unit cell) with four single-ended ports corresponding to half a CSRR-loaded differential microstrip line, as was derived in Fig. 4.16a

in [25] as an expanded work. The structure under study consists of a pair of parallel microstrip lines with an array of CSRRs symmetrically and periodically etched in the ground plane (see Fig. 5.1). The symmetry plane of the CSRRs, which behaves as a magnetic wall (Sect. 2.3.5), is aligned with the symmetry plane of the differential line that exhibits either a magnetic wall or an electric wall for the even and odd modes, respectively (Sect. 4.1).

The analysis of the structure was described in Sect. 4.2.2. An equivalent circuit model for the whole structure was proposed therein, and the circuits derived from even- and odd-mode excitations were also obtained. The resulting circuit for the odd mode is that of a conventional transmission line (with an all-pass response), while the circuit for the even mode is equivalent to that of a CSRR-loaded microstrip line (with a stopband response). The latter, important for this subsection, is reproduced in Fig. 5.2. For the considered symmetric CSRR-loaded differential lines, there is line-to-resonator magnetoelectric coupling. Despite that, mixed coupling is neglected and only the electric coupling is modeled. This approximation eases the design of common-mode suppressed differential lines, and is not crucial because mixed coupling is not relevant (in the considered topologies) for signal rejection.

There are three main strategies to enhance the common-mode rejection bandwidth in these CSRR-based structures: (*i*) to tune the resonance frequency of several resonators in order to cover a wide band; (*ii*) to couple the resonators as much as possible; and (*iii*) to enhance the bandwidth of the individual resonator-loaded transmission-line unit cells (and, obviously, a combination of all these approaches). Since the approach of multiple tuned resonators is efficient but many particles are needed if narrowband resonators are used, we focus on the other two approaches (i.e. employing identical and closely-spaced wideband resonant unit cells, periodically cascaded).

In order to gain insight into the rejection bandwidth for the common mode, the dispersion relation given by (3.4), namely,

$$\cosh(\gamma l) = \frac{1}{2}\left(A_{11} + A_{22} \pm \sqrt{(A_{11} - A_{22})^2 + 4A_{12}A_{21}}\right), \tag{5.1}$$

is an appropriate means. The elements of the $\boldsymbol{A}$ matrix in the circuit of Fig. 5.2 are

$$A_{11} = 1 - L_e C_e \omega^2/2, \tag{5.2a}$$

$$A_{12} = L_e C_e \omega^2/2, \tag{5.2b}$$

$$A_{21} = -\frac{C_e}{C_R}, \tag{5.2c}$$

$$A_{22} = 1 + \frac{C_e}{C_R} + \frac{C_c}{2C_R}\left(1 - \frac{1}{L_c C_c \omega^2}\right). \tag{5.2d}$$

As was discussed in Sect. 3.2.1, the inter-resonator electric coupling widens the stopband due to the presence of complex modes. Therefore, inter-resonator coupling helps to broaden the stopband of the common mode in CSRR-loaded differential lines, but bandwidth enhancement based on CSRR-to-CSRR coupling is rather limited. This makes apparent the convenience of implementing common-mode rejection filters with broadband resonators.

According to the equivalent circuit model, to widen the rejection bandwidth of an individual CSRR-based elemental cell, it is necessary to increase the line-to-resonator coupling capacitance, and to reduce the CSRR inductance and capacitance. In order to achieve a strong electric coupling between the pair of lines and the resonator, the lines must be fitted face-to-face with the CSRR inner region, and must be as wide as possible. Accordingly, loosely coupled lines are considered, since the width of the lines necessary to achieve a 50-Ω odd-mode impedance is wider (from Fig. 4.16 it can be drawn that coupling between the lines decreases the odd-mode characteristic impedance). To reduce the inductance and capacitance of the CSRR, it is necessary to increase the witdh and the separation of the slot rings. Obviously, this results in an electrically larger CSRR (for a given notch frequency) and the equivalent circuit model is no longer valid within a wide band (only in the vicinity of the notch), but the achievable rejection bandwidth is wider. As a final step, to enhance inter-resonator coupling, square-shaped and closely-spaced CSRRs are designed. Two order-3 CSRR-based prototypes, which will be discussed in the next two paragraphs, are reported in [24] on the basis of these design rules.

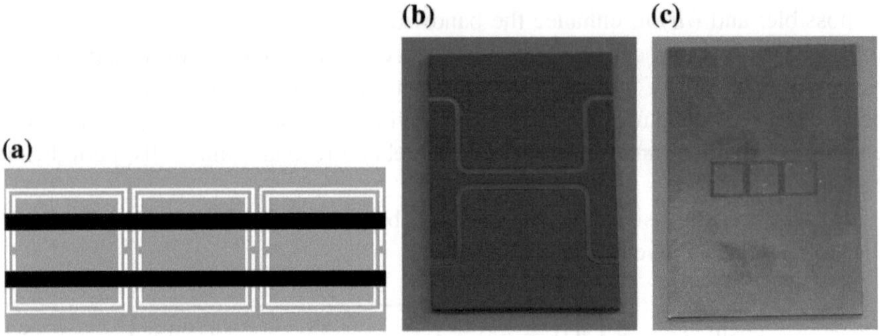

Fig. 5.3 **a** Layout, and **b** top and **c** bottom views of an order-3 CSRR-loaded differential microstrip line for common-mode suppression using electrically small unit cells. Dimensions are: $W = 1$ mm, $S = 2.5$ mm (50-Ω odd-mode line, and 57-Ω even-mode line), $c = d = 0.2$ mm, $l_1 = l_2 = 7.6$ mm, and $d_R = 0.15$ mm. The substrate is *Rogers RO3010* with thickness $h = 1.27$ mm, relative permittivity $\epsilon_r = 10.2$, and loss tangent $\tan\delta = 0.0023$. Access lines are introduced to measure the device

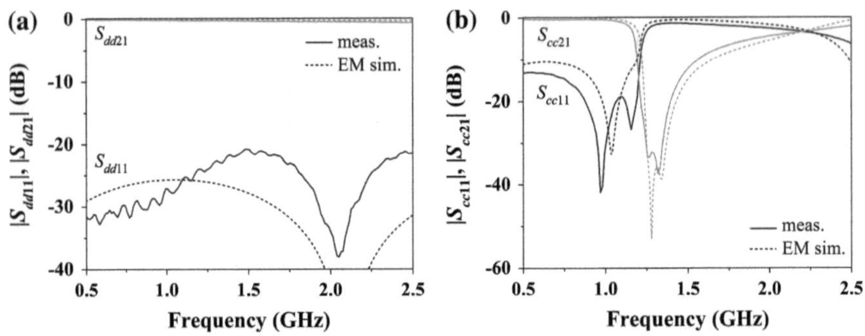

Fig. 5.4 **a** Differential-mode return loss and insertion loss ($|S_{dd11}|$ and $|S_{dd21}|$), and **b** common-mode return loss and insertion loss ($|S_{cc11}|$ and $|S_{cc21}|$) for the CSRR-based structure of Fig. 5.3

In one of the prototypes, shown in Fig. 5.3, the purpose is to reduce the size of the structure as much as possible. For this reason, a CSRR unit cell with narrow and closely-spaced rings is considered, guaranteeing the validity of the circuit model. Indeed, the synthesis of the filter is systematic, as will be explained later. The device is very small, since the dimensions of the slotted region of the structure are 23×7.6 mm ($0.28\lambda_e \times 0.09\lambda_e$, where λ_e is the guided wavelength at the central frequency for the even mode). Figure 5.4 plots the differential- and common-mode frequency response of the device. The measured insertion loss for the differential mode is smaller than 0.1 dB, but the common-mode rejection bandwidth at −20 dB (14 %) is not optimized in this structure (as mentioned before, wideband resonators are needed). Although within the stopband there are three resonance frequencies, since the two lowest resonance frequencies are very close to each other and losses round the transmission zeros off, these two transmission zeros degenerate to the same one.

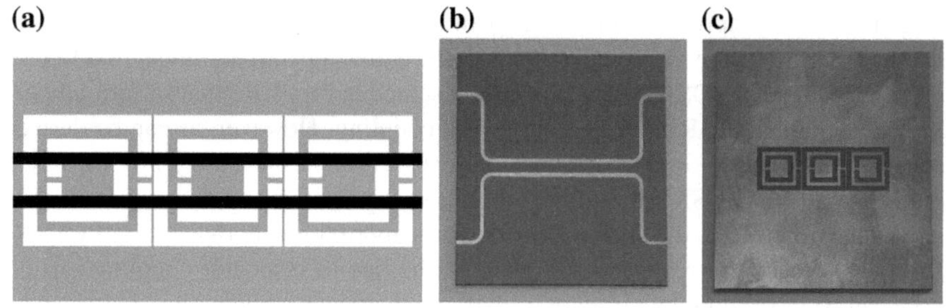

Fig. 5.5 **a** Layout, and **b** top and **c** bottom views of an order-3 CSRR-loaded differential microstrip line with electrically large unit cells. Dimensions are: $W = 1$ mm, $S = 2.5$ mm (50-Ω odd-mode line), $c = 1.2$ mm, $d = 0.8$ mm, $l_1 = l_2 = 10.8$ mm, and $d_R = 0.2$ mm. The substrate is *Rogers RO3010* with $h = 1.27$ mm, $\epsilon_r = 10.2$, and $\tan\delta = 0.0023$. Access lines are introduced to measure the device

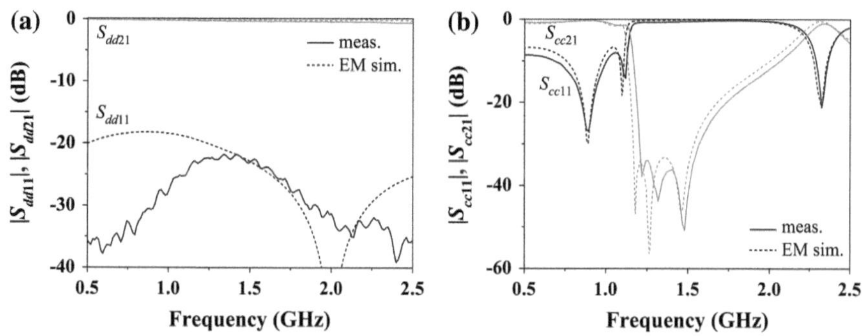

Fig. 5.6 **a** Differential- and **b** common-mode frequency response for the CSRR-based structure of Fig. 5.5. Reprinted with permission from [25]

The other design is aimed at enhancing the rejection bandwidth. For this purpose, CSRRs with wider rings and inter-ring separation are designed (Fig. 5.5). The model of the unit cell is not so simple in this case because the particle cannot be considered to be electrically small; therefore the optimization is directly made at layout level. The frequency response is plotted in Fig. 5.6. The device exhibits a stopband bandwidth for the common mode of 37 % (from 1.18 to 1.74 GHz) with more than 20 dB rejection. It is remarkable that the insertion loss for differential signals is smaller than 0.5 dB from DC to at least 2.5 GHz. The dimensions are 32.8 × 10.8 mm ($0.43\lambda_e \times 0.14\lambda_e$). It is therefore verified that the rejection bandwidth is broadened by enlarging the CSRR dimensions.

In the expanded work [25], as an alternative to the CSRR, the DS-CSRR is utilized as well. The DS-CSRR is bisymmetric, does not suffer from cross-polarization, and induces two orthogonal magnetic walls at the fundamental resonance (Sect. 2.3.5)

which make it appropriate for the selective mode suppression.[6] Under the assumption that the DS-CSRRs are electrically coupled, the equivalent circuit model of differential lines loaded with DS-CSRRs is that represented in Fig. 4.16, that is, formally the same as that using CSRRs neglecting mixed coupling. However, as can be derived from the relation SRR-to-DS-SRR and duality (Sects. 2.3.2 and 2.3.5), in comparison to a CSRR, the DS-CSRR has almost the same capacitance, but four times smaller inductance. Therefore, the size of DS-CSRRs is approximately twice that of CSRRs (for a given resonator resonance frequency), the line-to-resonator coupling capacitance can be larger and, by virtue of a smaller inductance, a wider bandwidth can be obtained (for a given transmission zero frequency).

An evaluation of the rejection bandwidth and size to compare the usefulness of CSRRs and DS-CSRRs is performed by electromagnetic simulations. The fractional bandwidth is taken as

$$FBW = \frac{BW}{f_c} = \frac{|f_1 - f_2|}{\sqrt{f_1 f_2}}, \tag{5.3}$$

where f_1 and f_2 are the lower and upper frequency limits of the stopband, respectively, and f_c is the filter central frequency. The results, for different orders and central filter frequencies (actually the transmission zero frequency of a single-cell structure, f_z), are shown in Fig. 5.7. As expected, the common-mode rejection bandwidth is wider for DS-CSRR-loaded differential lines. Though for a given central frequency an individual DS-CSRR is larger than a CSRR, if one compares CSRR- and DS-CSRR-loaded lines with the same area, it is seen that the rejection bandwidth is wider for the latter. With these results, it can be concluded that DS-CSRRs exhibit a good balance between filter size and achievable bandwidth.

From Fig. 5.7, it can be as well drawn that the electrical size of the structures increases with the central frequency. This occurs because when shifting the central frequency upward, the particle dimensions are not scaled down, that is, the resonator side length is decreased, but the ring width and inter-ring spacing (c and d) are kept constant (as well as the substrate height, h). It should also be noted that the common-mode rejection bandwidth narrows as the central frequency increases. This aspect is attributed to a drastic decrease in the coupling capacitance between the differential line and the resonators, owing to the fact that the substrate thickness is not scaled down.

Provided that the unit cell is small in terms of the wavelength, the circuit model is valid, and a systematic design procedure to determine the filter dimensions on the basis of the even-mode equivalent circuit model can be applied. Furthermore, the maximum achievable rejection bandwidth (by an infinitely periodic array of cells) can be predicted through the dispersion relation. Note that since the DS-CSRR is electrically larger than the CSRR, it is expected that the common-mode circuit model provides an accurate description over a wider frequency band for CSRR-loaded lines. This means that any prediction based on the circuit model is expected to be more

---

[6] There are two possible alignments. Particularly, the slits of the inner rings are lined up with the line axis, as shown in Fig. 5.9a. This orientation is aimed at obtaining a coupling dominantly electric through a chain of DS-CSRRs.

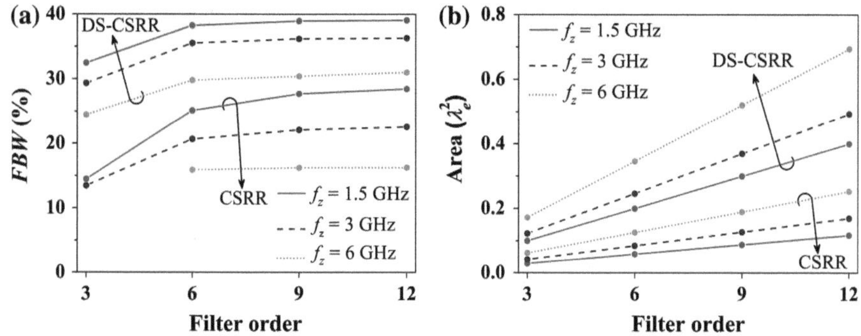

Fig. 5.7 **a** Fractional rejection bandwidth at −20 dB for the common mode given by lossless electromagnetic simulations and **b** area as a function of the even-mode guided wavelength λ_e of CSRR- and DS-CSRR-loaded differential lines. Dimensions are: $l_1 = l_2$, $c = d = 0.2$ mm, and $d_R = 0.2$ mm; for the CSRRs, $l_1(6\text{ GHz}) = 7.3$ mm, $l_1(3\text{ GHz}) = 4.3$ mm, and $l_1(6\text{ GHz}) = 2.6$ mm; for the DS-CSRRs, $l_1(6\text{ GHz}) = 13.8$ mm, $l_1(6\text{ GHz}) = 7.5$ mm, and $l_1(6\text{ GHz}) = 4.3$ mm; for the differential line, $2W + S = l_1 - 2(2c - d) + 0.4$ mm (50-Ω odd-mode characteristic impedance). The substrate is *Rogers RO3010* with $h = 1.27$ mm and $\epsilon_r = 10.2$. Reprinted with permission from [25]

accurate for CSRR-loaded lines.[7] Nevertheless, the advantage of using DS-CSRRs is that the bandwidth can be enhanced while preserving (to some extent) the description of the particle through an *LC* resonant tank.

To synthesize the common-mode filter, the filter dimensions are perfectly determined using the following procedure. By setting narrow *c* and *d* (as compared to the resonator side length), the unit cell is electrically small and the circuit model can be applied in filter synthesis. Following the design rules previously explained, the only design parameter is the side length of the square resonator. The side length of the resonator is determined from the model of the CSRR reported in [26] (which gives L_c and C_c for circular resonators) and the capacitance of the coupled lines for the even mode (which gives C_e). The transmission zero frequency, in the absence of inter-resonator coupling, that is[8]

$$f_z = \frac{1}{2\pi\sqrt{2L_c(C_e + C_c/2)}}, \tag{5.4}$$

is tuned at the required central filter frequency, $f_z = f_c$.[9] This provides an estimated CSRR side length; for a DS-CSRR the side length can be determined by taking into account that the inductance is four times smaller than the inductance of the CSRR.

[7] Note, however, that the circuit model for the CSRR is approximate, since the line-to-resonator magnetic coupling is ignored.

[8] Using CSRRs, the transmission zero frequency also depends on the line-to-resonator magnetic coupling, but its contribution is neglected.

[9] The transmission zero frequency for an isolated unit cell provides a reasonable estimate of the central filter frequency.

Table 5.1 Extracted circuit parameters and maximum fractional bandwidth inferred from the dispersion relation of the circuit model for the **a** CSRR- and **b** DS-CSRR-loaded differential lines in Fig. 5.7

f_z (GHz)	L_e (nH)	C_e (pF)	L_C (nH)	C_C (pF)	C_R (pF)	FBW_{max} (%)
(a)						
1.5	6.3	1.1	2.1	3.2	0.1	30.7
3	4.2	0.5	1.0	2.0	0.06	26.4
6	3.0	0.2	0.4	1.4	0.08	20.7
(b)						
1.5	15.9	1.6	1.0	8.6	0.64	32.6
3	9.0	0.8	0.45	4.8	0.18	29.4
6	6.9	0.3	0.15	3.6	0.1	23.6

The circuit parameters are extracted as follows. (*i*) The parameters of a one-cell structure (L_e, C_e, L_c and C_c) are extracted according to the procedure reported in [27] (due to line-to-resonator magnetoelectric coupling, it is essential to properly choose the input and output ports; otherwise the circuit parameter extraction fails). (*ii*) The inter-resonator coupling capacitance (C_R) is considered to be an adjustable parameter to fit the electromagnetic simulation of a two-cell structure. For simplicity, C_c is not readjusted by the effects of inter-resonator coupling

Next, optimization of the resonator side length in order to fit the required transmission zero is necessary since the considered resonators are square-shaped.

For the purpose of predicting the maximum achievable bandwidth, i.e., the bandwidth obtained by considering an infinite number of cells, the dispersion relation is used. However, since optimization at layout level is required, it is necessary to extract the parameters of the circuit model. Once the circuit parameters are known, expression (5.1) can be evaluated, and the common-mode stopband bandwidth can be determined. Following this procedure, the maximum achievable rejection bandwidth is obtained for the CSRR and DS-CSRR common-mode filters in Fig. 5.7. The extracted circuit parameters and estimated fractional bandwidths are listed in Table 5.1. By comparing the maximum fractional bandwidths predicted by the reported approach with the saturation values in Fig. 5.7,[10] it is observed that the reported approach is more accurate for CSRR-loaded lines, as expected.

Figure 5.8 compares the circuit and electromagnetic simulations (common mode) of the order 1, 2 and 3 CSRR-based common-mode filters designed to exhibit f_z = 1.5 GHz.[11] The agreement between the circuit and electromagnetic simulations is reasonably good, bearing in mind that mixed coupling is neglected. This comparison demonstrates that mixed coupling can be neglected, as a first-order approximation, for attenuation purposes.

[10]This comparison is meaningful for a large number of cells. A discussion on whether the bandwidth inferred from the dispersion relation (FBW_{max}) is related to the −20 dB bandwidth of a structure with a large number of cells (FBW) is given in Appendix E.

[11]As explained in Sect. 3.2.1, the inter-resonator capacitance at input and output ports are left opened, and the equivalent circuit reduces to a two-port network.

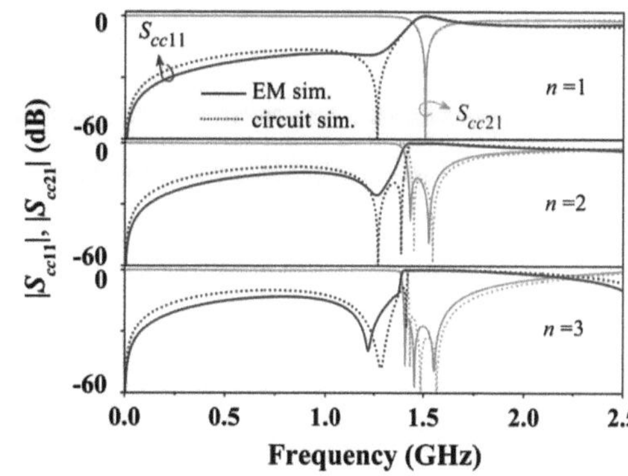

Fig. 5.8 Common-mode return loss and insertion loss given by electromagnetic and circuit simulations for the order 1, 2 and 3 common-mode filters based on CSRRs for $f_z = 1.5$ GHz in Fig. 5.7. Circuit parameters are indicated in Table 5.1a. Reprinted with permission from [25]

It is noticed from Fig. 5.7 that six resonators are sufficient to nearly achieve the maximum rejection bandwidth. Thus, following a systematic approach based on the circuit model of the common mode, the estimation of the maximum rejection bandwidth allows us to infer whether a specified rejection bandwidth and central frequency can be roughly fulfilled or not.[12] If the required bandwidth is wider, we are forced to consider electrically large resonators (the circuit model cannot be used) and/or multiple tuned resonators. In these cases, however, filter design and maximum bandwidth estimation are not so straightforward.

In order to illustrate the design methodology and the prediction of the maximum bandwidth, we consider a DS-CSRR with small ring width and separation (obviously, DS-CSRRs with wide and distant rings can also be considered to further enhance the common-mode rejection bandwidth, but in this case the circuit model cannot be used for synthesis purposes). To compare with the CSRR-based structure in Fig. 5.5 [24], the target is to implement a common-mode filter roughly centered at 1.35 GHz and exhibiting at least 35 % fractional rejection bandwidth. According to the previous methodology, these specifications cannot be fulfilled by using CSRRs with $c = d = 0.2$ mm. Conversely, it is possible to achieve these filter requirements by means of DS-CSRRs. Indeed, the estimated maximum FBW is found to be 37.3 %, but we do expect a larger value since the model tends to slightly underestimate the maximum achievable bandwidth for DS-CSRR-loaded lines (this can be appreciated by comparing the bandwidths in Fig. 5.7 and Table 5.1). For comparison purposes, a rectangular-shaped DS-CSRR is considered with its transverse side length, l_2, identical to that of the CSRR reported in [24]. This, moreover, favors the electric coupling between the pair of lines and the DS-CSRRs, and hence the common-mode stopband expansion (the reason is that the DS-CSRR longitudinal side is longer than the transverse one, and this increases the coupling capacitance C_e as compared to that of a square-shaped DS-CSRR with identical transmission zero frequency). The longitudinal side length, l_1, is thus the single design parameter, and is determined following the same approach applied to square-shaped particles.

[12] The reported approach allows us to infer the maximum achievable common-mode rejection bandwidth, rather than specifying the bandwidth to a particular value.

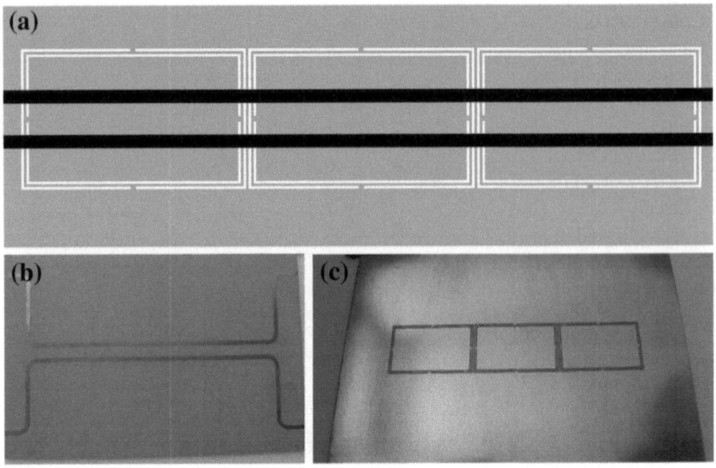

Fig. 5.9 **a** Layout and photograph of the **b** top and **c** bottom views of the designed differential line with common-mode suppression based on DS-CSRRs. Dimensions are: $W = 1$ mm, $S = 2.5$ mm (50-Ω odd-mode line), $c = d = 0.2$ mm, $l_1 = 17.6$ mm, $l_2 = 10.8$ mm, and $d_R = 0.2$ mm. The substrate is *Rogers RO3010* with $h = 1.27$ mm, $\epsilon_r = 10.2$, and $\tan\delta = 0.0023$

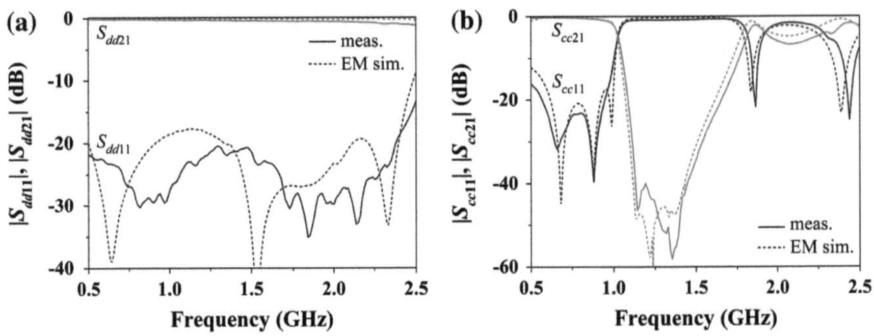

Fig. 5.10 **a** Differential- and **b** common-mode frequency response for the DS-CSRR-based structure of Fig. 5.9. Reprinted with permission from [25]

The photograph and frequency response of the device, an order-3 common-mode filter similar to that reported in [24] using CSRRs, are shown in Figs. 5.9 and 5.10, respectively (this filter order is found to be enough to satisfy the bandwidth requirements). As can be noticed, the differential signal is almost unaltered whilst the common mode is rejected more than 20 dB within a 41 % fractional bandwidth (i.e. comparable to that achieved using CSRRs with wide and widely spaced rings). Like in the CSRR-based filter, the measured insertion loss is smaller than 0.5 dB (at least up to 2.5 GHz). The DS-CSRR-based structure is a bit larger ($0.64\lambda_e \times 0.14\lambda_e$) than that utilizing CSRRs, but the design is carried out following the reported systematic

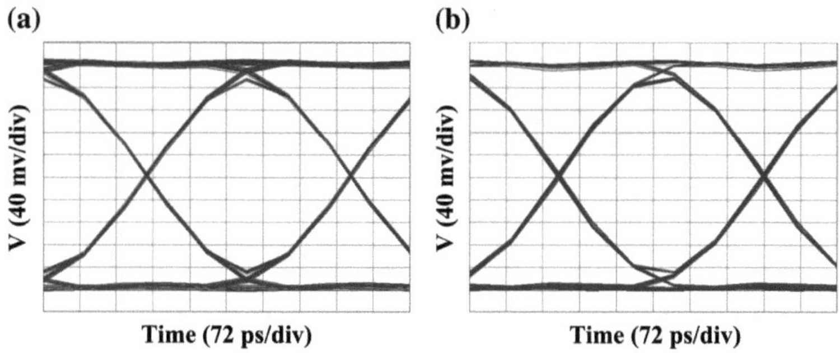

Fig. 5.11 Measured differential eye diagrams for the differential line of Fig. 5.9 **a** with and **b** without DS-CSRRs. An excitation of 0.2 V amplitude in 2.5 Gb/s is applied. Reprinted with permission from [25]

Table 5.2 Measured eye parameters of the DS-CSRR-based common-mode filter

	With DS-CSRRs	Without DS-CSRRs
Eye height	278 mV	281 mV
Eye width	371 ps	383 ps
Jitter (PP)	29.3 ps	16.9 ps
Eye opening factor	0.76	0.76

procedure.[13] Eye diagrams of the differential line with and without DS-CSRRs are also measured, and are plotted in Fig. 5.11. The eye diagram quality in terms of some parameters is compared for these two structures (see Table 5.2). The eye diagrams reveal that the common-mode rejection does not degrade significantly the differential signals. The peak-to-peak jitter varies notably, but it is still within very acceptable limits for the DS-CSRR-based structure. Moreover, the eye opening factor, which measures the ratio of eye height and eye amplitude, is identical. In view of these results, the differential-mode signal integrity is almost unaltered by the presence of the DS-CSRRs. The common-mode suppression in differential microstrip lines loaded with DS-CSRRs is hence experimentally verified.

Additionally, another potential application is presented. Specifically, the common-mode suppressed band is enhanced in a balanced passband filter consisting of a pair of coupled folded stepped impedance resonators (FSIRs) fed by a differential line (see Fig. 5.12). The filter by itself rejects the common mode due to the symmetry of the structure, since the symmetry plane of the resonator behaves as an electric wall at the first resonance. Therefore, such resonators cannot be excited by means of common-mode signals, and the even mode is reflected back to the source due to the

[13]For a particular transmission zero frequency, the accuracy of the circuit model does not depend exclusively on the physical size, but also on the ring width and inter-ring distance. It is for this reason that, despite the fact that the DS-CSRR is physically larger than the CSRR, the circuit model is more accurate for the former.

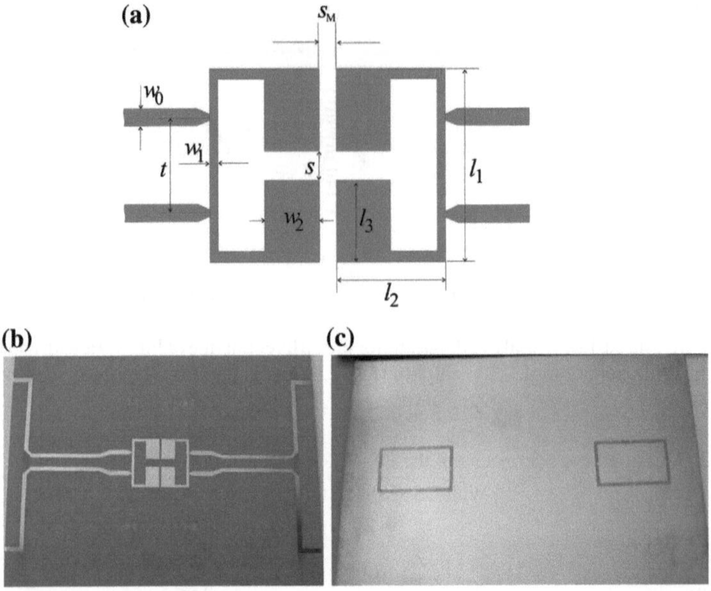

Fig. 5.12 **a** Layout, and photographs of the **b** top and **c** bottom views of the designed differential bandpass filter with improved common-mode rejection. Dimensions are: for the DS-CSRRs, $c = d = 0.2$ mm, $l_1 = 17.6$ mm, $l_2 = 10.8$ mm; for the differential line, $W = 1$ mm, and $S = 2.5$ mm (50-Ω odd-mode line); for the differential filter, $S_M = 0.5$ mm, $s = 2$ mm, $w_0 = 1.2$ mm, $w_1 = 0.7$ mm, $w_2 = 3.7$ mm, $l_1 = 12.8$ mm, $l_2 = 7.5$ mm, $l_3 = 5.4$ mm, and $t = 5.5$ mm. The substrate is *Rogers RO3010* with $h = 1.27$ mm, $\epsilon_r = 10.2$, and $\tan\delta = 0.0023$

gaps between the pair of FSIRs. However, the rejection level of the common mode in the region of interest is very limited since it depends on the distance between resonators and such inter-resonator distance is dictated by filter specifications.[14] By merely cascading two identical DS-CSRRs (whose common-mode rejection band is generated in the passband region of the differential filter) to the proposed balanced filter, the common-mode rejection is roughly increased from 20 to 50 dB in the region of interest (see Fig. 5.13). On the other hand, the filter response (differential mode) is practically unaffected by the presence of the DS-CSRRs. To completely characterize the differential filter, the group delay for the differential mode is also measured, and shown in Fig. 5.14. All the previous results indicate that DS-CSRRs provide an efficient path to enhance the common-mode noise rejection in balanced filters.

Finally, the presented CSRR- and DS-CSRR-based differential lines with common-mode suppression are compared with other structures reported in the literature in Table 5.3. The filter reported in [4] is very competitive in terms of size and bandwidth, but the maximum rejection level is moderate. Moreover, such common-

[14]The proposed differential filter is a second-order Chebyshev bandpass filter with a central frequency of 1.37 GHz, a fractional bandwidth of 10 %, and 0.1 dB ripple. The design of the filter is done following the procedure described in [28].

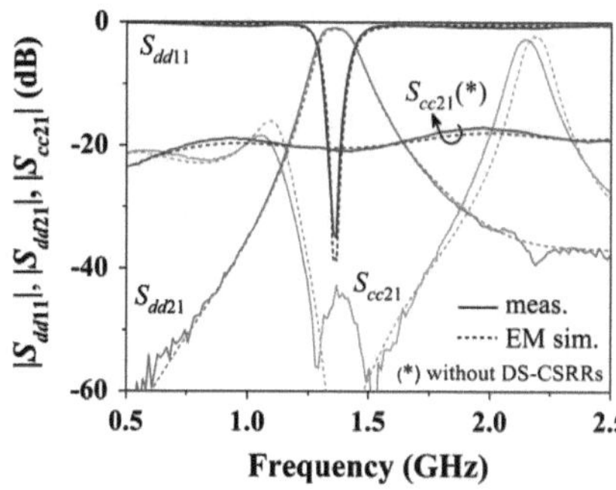

Fig. 5.13 Simulated and measured differential-mode return loss and insertion loss ($|S_{dd11}|$ and $|S_{dd21}|$), and common-mode insertion loss ($|S_{cc21}|$) for the balanced bandpass filter of Fig. 5.12 with and without the presence of the DS-CSRRs. Reprinted with permission from [25]

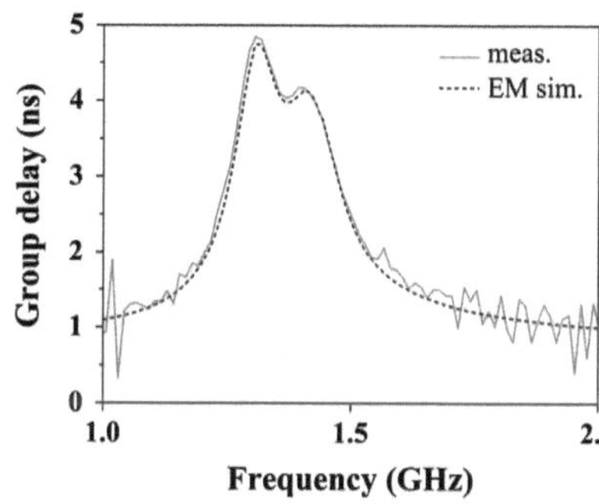

Fig. 5.14 Group delay for the differential filter of Fig. 5.12. Reprinted with permission from [25]

mode filter is based on a multilayer structure and this has extra cost and complexity. The filters reported in [7, 8] have larger dimensions than the CSRR- and DS-CSRR-based filters. The structures reported in [8, 10] exhibit a huge stopband considering a −10 dB bandwidth, but the maximum rejection level is poor as compared to that

Table 5.3 Comparison of several differential lines with common-mode suppression, including the structures based on CSRRs in Fig. 5.5 [24] and on DS-CSRRs in Fig. 5.9 [25]

Reference	Length/λ_e	Width/λ_e	Area/λ_e^2	−10 dB $FBW(\%)$	−20 dB $FBW(\%)$	−40 dB $FBW(\%)$
Tsai and Wu [4]	0.26	0.16	0.04	60	32	–
Liu et al. [7]	0.76	0.47	0.36	70	53	–
Wu, Tsai and Wu [8]	0.44	0.44	0.19	87	–	–
Fernández-Prieto et al. [10]	0.32	0.22	0.07	106	67	–
Naqui et al. [24]	0.43	0.14	0.06	54	37	14
Naqui et al. [25]	0.64	0.14	0.08	51	41	23

The dimensions correspond to the active area (slotted area in the defected ground structures)

of the proposed filters. The common-mode filter presented in [10] is especially competitive, i.e., it exhibits a very good combination of dimensions and bandwidth, although the common-mode rejection level is lower (it does not reach 40 dB) than those achievable with periodically cascaded CSRRs and DS-CSRRs. In summary, in the reported approach, the combination of size, rejection bandwidth, and rejection level is competitive. The proposed common-mode suppression strategy is technologically simple and low cost, the systematic design procedure is simple, and the limits on the maximum rejection bandwidth can be easily inferred.

5.3.2 ELC- and MLC-Loaded Differential Microstrip Lines

Section 4.2.3 was addressed to analyze the effect of electric and magnetic inductive-capacitive (ELC and MLC) resonators when they load symmetrically a differential microstrip line [29, 30].

The advantage of the ELC (against the CSRR, DS-CSRR, and MLC) to suppress the common mode is that the ground plane is kept unaltered (this is frequently a requirement in certain applications). However, the preliminary results in Figs. 4.17a and 4.18b and those in Figs. 4.20b and 4.26b seem to indicate that the even mode is suppressed in a wide band using neither ELC nor MLC resonators. Since the rejection is not as efficient as by means of CSRRs, no common-mode suppressed differential line employing ELC or MLC resonators has been designed in this thesis.

References

1. T.C. Edwards, M.B. Steer, *Foundations of Interconnect and Microstrip Design.* (Wiley, New York, 2000)
2. C.R. Paul, *Introduction to Electromagnetic Compatibility*, 2nd edn. (Wiley, New York, 2006)
3. D. Bockelman, W. Eisenstadt, Combined differential and common-mode scattering parameters: theory and simulation. IEEE Trans. Microw. Theory Tech. **43**(7), 1530–1539 (1995)
4. C.-H. Tsai, T.-L. Wu, A broadband and miniaturized common-mode filter for gigahertz differential signals based on negative-permittivity metamaterials. IEEE Trans. Microw. Theory Tech. **58**(1), 195–202 (2010)
5. C.-H. Wu, C.-H. Wang, C.H. Chen, Novel balanced coupled-line bandpass filters with common-mode noise suppression. IEEE Trans. Microw. Theory Tech. **55**(2), 287–295 (2007)
6. B.-C. Tseng, L.-K. Wu, Design of miniaturized common-mode filter by multilayer low-temperature co-fired ceramic. IEEE Trans. Electromagn. Compat. **46**(4), 571–579 (2004)
7. W.-T. Liu, C.-H. Tsai, T.-W. Han, T.-L. Wu, An embedded common-mode suppression filter for GHz differential signals using periodic defected ground plane. IEEE Microw. Wireless Compon. Lett. **18**(4), 248–250 (2008)
8. T.-L. Wu, C.-H. Tsai, T.-L. Wu, T. Itoh, A novel wideband common-mode suppression filter for gigahertz differential signals using coupled patterned ground structure. IEEE Trans. Microw. Theory Tech. **57**(4), 848–855 (2009)
9. A. Fernández-Prieto, J. Martel, F. Medina, F. Mesa, S. Qian, J. S. Hong, Microstrip/CPW differential lines for common-mode suppression, in *Metamaterials '2011: 5th International*

Congress on Advanced Electromagnetic Materials Microwaves and Optics, (Barcelona, Spain, 2011)
10. A. Fernández-Prieto, J. Martel, F. Medina, F. Mesa, S. Qian, J.-S. Hong, J. Naqui, F. Martín, Dual-band differential filter using broadband common-mode rejection artificial transmission line. Prog. Electromagn. Res. **139**, 779–797 (2013)
11. T.-W. Weng, C.-H. Tsai, C.-H. Chen, D.-H. Han, T.-L. Wu, Synthesis model and design of a common-mode bandstop filter (CM-BSF) with an all-pass characteristic for high-speed differential signals. IEEE Trans. Microw. Theory Tech. **62**(8), 1647–1656 (2014)
12. J. Choi, P. Hon, T. Itoh, Dispersion analysis and design of planar electromagnetic bandgap ground plane for broadband common-mode suppression. *IEEE Microw. Wireless Compon. Lett.* (2014)
13. T.-L. Wu, F. Buesink, F. Canavero, Overview of signal integrity and EMC design technologies on PCB: fundamentals and latest progress. IEEE Trans. Electromagn. Compat. **55**(4), 624–638 (2013)
14. C.-H. Wu, C.-H. Wang, C.H. Chen, Stopband-extended balanced bandpass filter using coupled stepped-impedance resonators. IEEE Microw. Wireless Compon. Lett. **17**(7), 507–509 (2007)
15. C.-H. Lee, C.-I. Hsu, C.-C. Hsu, Balanced dual-band BPF with stub-loaded SIRs for common-mode suppression. IEEE Microw. Wireless Compon. Lett. **20**(2), 70–72 (2010)
16. J. Shi, Q. Xue, Dual-band and wide-stopband single-band balanced bandpass filters with high selectivity and common-mode suppression. IEEE Trans. Microw. Theory Tech. **58**(8), 2204–2212 (2010)
17. P. Vélez, J. Naqui, A. Fernández-Prieto, M. Durán-Sindreu, J. Bonache, J. Martel, F. Medina, F. Martín, Differential bandpass filter with common-mode suppression based on open split ring resonators and open complementary split ring resonators. IEEE Microw. Wireless Compon. Lett. **23**(1), 22–24 (2013)
18. P. Vélez, J. Naqui, A. Fernández-Prieto, M. Durán-Sindreu, J. Bonache, J. Martel, F. Medina, F. Martín, Differential bandpass filters with common-mode suppression based on stepped impedance resonators (SIRs), in *IEEE MTT-S International Microwave Symposium*, Seattle, (WA, USA, 2013)
19. P. Vélez, M. Durán-Sindreu, J. Naqui, J. Bonache, F. Martín, Common-mode suppressed differential bandpass filter based on open complementary split ring resonators fabricated in microstrip technology without ground plane etching. Microw. Opt. Technol. Lett. **56**(4), 910–916 (2014)
20. P. Vélez, M. Durán-Sindreu, A. Fernández-Prieto, J. Bonache, F. Medina, F. Martín, Compact dual-band differential power splitter with common-mode suppression and filtering capability based on differential-mode composite right/left-handed transmission-line metamaterials. IEEE Antennas Wireless Propag. Lett. **13**, 536–539 (2014)
21. J. Zhou, Y.-C. Chiang, W. Che, Compact wideband balanced bandpass filter with high common-mode suppression based on cascade parallel coupled lines. IET Microw. Antennas Propag. **8**(8), 564–570 (2014)
22. P. Vélez, J. Naqui, A. Fernández-Prieto, J. Bonache, J. Mata-Contreras, J. Martel, F. Medina, F. Martín, Ultra-compact (80 mm^2) differential-mode ultra-wideband (UWB) bandpass filters with common-mode noise suppression. IEEE Trans. Microw. Theory Tech. **63**(4), 1272–1280 (2015)
23. R.N. Simons, *Coplanar Waveguide Circuits, Components, and Systems*. (Wiley, New York, 2001)
24. J. Naqui, A. Fernández-Prieto, M. Durán-Sindreu, J. Selga, F. Medina, F. Mesa, F. Martín, Split rings-based differential transmission lines with common-mode suppression, in *IEEE MTT-S International Microwave Symposium Digest* (Baltimore, MD, USA, 2011)
25. J. Naqui, A. Fernández-Prieto, M. Durán-Sindreu, F. Mesa, J. Martel, F. Medina, F. Martín, Common-mode suppression in microstrip differential lines by means of complementary split ring resonators: Theory and applications. IEEE Trans. Microw. Theory Tech. **60**(10), 3023–3034 (2012)
26. J.D. Baena, J. Bonache, F. Martín, R.M. Sillero, F. Falcone, T. Lopetegi, M.A.G. Laso, J. García-García, I. Gil, M.F. Portillo, M. Sorolla, Equivalent-circuit models for split-ring resonators and

complementary split-ring resonators coupled to planar transmission lines. IEEE Trans. Microw. Theory Tech. **53**(4), 1451–1461 (2005)

27. J. Bonache, M. Gil, I. Gil, J. García-García, F. Martín, On the electrical characteristics of complementary metamaterial resonators. IEEE Microw. Wireless Compon. Lett. **16**(10), 543–545 (2006)
28. J.-S. Hong, M.J. Lancaster, *Microstrip Filters for RF/Microwave Applications*. (Wiley, New York, 2001)
29. J. Naqui, M. Durán-Sindreu, F. Martín, Selective mode suppression in microstrip differential lines by means of electric-LC (ELC) and magnetic-LC (MLC) resonators, in *4th International Conference on Metamaterials, Photonic Crystals and Plasmonics (META)*, (Sharjah, UAE, 2013)
30. J. Naqui, M. Durán-Sindreu, F. Martín, Selective mode suppression in microstrip differential lines by means of electric-LC (ELC) and magnetic-LC (MLC) resonators. Appl. Phys. A **115**, 637–643 (2014)

Chapter 6
Application of Symmetry Properties to Microwave Sensors

The present chapter is devoted to the application of the symmetry properties in resonator-loaded transmission lines described in Chap. 4 to microwave sensors. Specifically, symmetry rupture in transmission lines loaded with either single or pairs of resonators is exploited as a sensing mechanism. The novel proposed sensors may be classified into two categories in terms of the nature of the measurable physical quantity. Most of the presented sensors (all except one) are spatial sensors that involve the measurement of spatial variables (e.g. displacement or velocity). The other sensor that is proposed is a permittivity (an electromagnetic variable) sensor for differential measurements of the relative permittivity of materials. The former sensors are implemented using a transmission line loaded with a single symmetric resonator (where the principle of operation is based on a coupling-modulated resonance), whereas the latter sensor uses a transmission line loaded with a pair of symmetrically etched resonators (where resonance frequency splitting/shifting is the working principle).

Section 6.1 introduces microwave sensors (mainly resonator-based sensors), and some fundamental sensor parameters. In Sect. 6.2, the two reported sensing mechanisms are described depending on whether the line is loaded with either a single or a pair of resonators. Next sections are organized according to the measurand. Spatial sensors are addressed in Sects. 6.3–6.6, while the permittivity sensor is presented in Sect. 6.7.

6.1 Introduction

A *sensor* is a device that measures a physical quantity (called *measurand*) producing an electrical signal that can be read by an observer or an instrument. *Microwave sensors*, and particularly *metamaterial-inspired microwave sensors* [1], are attracting growing interest because they feature properties (e.g. miniaturization or cost) that cannot be provided by sensors working at lower frequencies or by other

J. Naqui, *Symmetry Properties in Transmission Lines Loaded with Electrically Small Resonators*, Springer Theses,
DOI 10.1007/978-3-319-24566-9_6

sensing principles. Microwave sensors are based on electromagnetic sensing principles, namely, a physical variable is measured from the variations of the characteristics of an RF/microwave signal. In recent years, there is a trend to integrate wireless sensors with RFID (radio-frequency identification) [2].[1] Hence, the combination of sensing and identification capabilities can be achieved by electromagnetic principles.

Resonator-based microwave sensors are devices where sensing is typically achieved through the variations experienced by the resonance frequency, phase, and/or quality factor of resonant structures. In particular, split-ring resonators (SRRs) and other related particles have been exhaustively used for sensing purposes due to their small electrical size and high-Q factor at the fundamental resonance. Different resonator-based sensors have been proposed in the literature, such as sensors for displacements [2–6], for organic tissue analysis [7–9], for microfluidic dielectric characterization [10–13], for crack detection [14], or for thickness measurements [15]. It is also worth mentioning that microwave sensors based on transmission-line elements have also been investigated [1, 16, 17].

A full description of sensor parameters may be found in many textbooks [18, 19], so that in this section the relevant parameters of the presented sensors in this thesis are briefly summarized:

- The relationship between an input physical variable and an output electrical variable is given by the so-called *transfer function*.
- The *input dynamic range* is the range of values of the measurand that can be measured.
- The *output dynamic range* corresponds to the range of values of the electrical variable.
- The *resolution* is the smallest increment of the input variable which can be sensed.
- The *sensitivity* is the slope of the transfer function and should be, in general, as high as possible and as constant as possible (ideal sensors are designed to be sensitive to the measurand, linearly or linearly to some simple mathematical function, e.g. logarithmic or exponential).
- The *cross-sensitivity* defines the sensitivity to other physical quantities, as generally occurs, degrading the sensor performance.
- The *linearity* indicates the deviation of the transfer function from an ideal straight line (a linear transfer function is characterized by a constant sensitivity).
- The *accuracy* defines the highest difference between the true and the measured value (i.e. defines the systemic error which measures an offset in a group of measurements).
- The *precision* of the measurement defines the random error (sometimes referred to as noise), where systematic errors are assumed to be removed, and therefore describes the dispersion of a set of measurements (i.e. quantifies the size of a group of measurements).

[1] Wireless sensors may be involved in many applications. For instance, industrial processes are driven and controlled by sensor-equipped devices, often autonomous and wirelessly interconnected [2].

Particularly, cross-sensitivity to environmental conditions (e.g. temperature, moisture, or pressure) are very common. Accordingly, environmental factors may and usually do affect the performance of sensors. Therefore, stability against variations of these conditions is often a very important requirement [18]. In the case of microwave sensors based on the frequency shift of resonant elements, their transfer functions can be altered by drifts caused by environmental changes.[2] As will be shown in the next section, the proposed sensors in this thesis intend to face with cross-sensitivities due to environmental conditions.

To end this section, it is important to discuss the measurement system. Typically, in resonator-based sensors, the sensing structure is merely a passive loaded transmission line whose cost is really insignificant. However, the physical magnitude is sensed by monitoring the transmission or reflection coefficient. Prototypes are usually characterized by measurements carried out by a vector network analyzer (VNA), which is a multifunctional, but heavy and expensive equipment. In a practical sensor, by contrast, the feeding and readout circuits attached to the passive sensing structure should be implemented by low-cost circuits and components [21]. Otherwise, the proposed sensing solutions cannot be regarded as low-cost sensor systems at all. In this regard, the lower the frequency of operation, the higher the availability of microwave circuits and components, and the lower the cost of the measurement circuitry.

6.2 Symmetry-Based Sensing

In this thesis two novel microwave sensor concepts inspired by symmetry properties of transmission lines loaded with resonant elements have been introduced. The sensing proposals use single or pairs of resonators, whose principles of operation are different. The former sensing strategy is exploited for the design of spatial sensors, whereas the latter for the implementation of permittivity sensors.[3] Most of the contributions of this thesis to microwave sensors is based on the former.

6.2.1 Coupling-Modulated Resonance

One of the sensing proposals is based on transmission lines loaded with a single symmetric resonator, where a coupling-modulated resonance can be controlled. Employing these structures, this thesis deals with the measurement of spatial

[2]For example, environmental factors may degrade the performance of displacement and alignment sensors based on transmission lines loaded with split-ring resonators where the operation principle is based on the shift of the resonance frequency [3]. On the other hand, since the resonance frequency in resonator-based sensors changes as the permittivity of the host medium changes in response to variations in environmental parameters, sensing of environmental conditions is indeed possible [20].

[3]Permittivity sensors based on a single resonator are also possible. Spatial sensors based on pairs of resonators can also be designed and, in fact, have been already reported in the literature [22]. This thesis, however, does not consider these sensors.

variables (e.g. alignment, displacement, or velocity). In these spatial sensors, the resonator must be attached to a substrate (or object) different from that of the transmission line in order to allow for a relative motion between the resonator and the line.[4]

The theory behind these sensors relies on properties derived from the misalignment of symmetry planes described in Sect. 4.3. If the resonator is properly aligned with the transmission line, line-to-resonator coupling is prevented, and the structure exhibits total transmission. By contrast, by breaking symmetry through the effects of a spatial variable, a transmission zero emerges, and the spatial variable may be inferred. The notched response is modulated by the net line-to-resonator coupling, which in turn is sensitive to the position of the resonator or, equivalently, to the level of asymmetry. Hence, the relative position between the two elements is measured through the transmission or reflection characteristics in the vicinity of resonance. It should be noted that, instead of looking at the line-to-resonator coupling, it is much more convenient to inspect the transmission or reflection coefficients since they can be directly measured at microwave frequencies.

The key advantage of the proposed sensing solution is its robustness against variable environmental conditions in virtue of the fact that the principle of operation relies on geometrical alignment/misalignment which cannot be altered by environmental factors.[5] The proposed approach is therefore especially suitable for alignment purposes. The coupling-modulated resonance can be viewed as a differential technique, since the net coupling can be expressed in terms of the difference between partial couplings related to each half of the structure.

Typically, amplitude and resonance frequency are characterized as sensing variables, although phase measurement may also be performed. However, most of the presented spatial sensors are characterized only by the notch depth, in logarithmic scale as usual to quantify attenuation.[6] Amplitude has been proven (Sect. 6.5.2) to exhibit better cross-sensitivities than resonance frequency to variations on the permittivity of the substrates, which may be produced by changing ambient factors.[7] On the other hand, a common disadvantage of sensing the amplitude is its susceptibility to electromagnetic interference (EMI) and noise that may affect seriously the performance of the sensors. Nevertheless, in the case that interfering signals are present, the frequency of operation can be easily shifted (e.g. by tailoring the

[4]Note that the resonator must be coupled to the line, but not connected to it. For instance, a folded stepped-impedance resonator (FSIR) may be used, but not a stepped-impedance shunt-stub (SISS).

[5]Though temperature can change the geometry by expansion of the materials, it is assumed that temperature does not change symmetry since microwave substrates exhibit a coefficient of thermal expansion (CTE) that allows for excellent dimensional stabilities.

[6]Note that although in most practical circuits and applications losses are undesirable, we take advantage of them to use the notch magnitude as a sensing electrical variable (if losses were absent, the notch magnitude would be always infinite). Losses, however, should be as low as possible to enhance the dynamic range of attenuation.

[7]Similarly, the notch depth is expected to exhibit low cross-sensitivities to fabrication tolerances in the nominal values of the dimensions (Sect. 6.5.2). By contrast, the resonance frequency of resonator-loaded lines is generally very sensitive to printing tolerances. For example, in ELC-loaded lines, cross-sensitivities affecting the resonance frequency and the notch magnitude can be analytically evaluated by (4.1) and (4.12), respectively.

resonator dimensions). Another way of reducing unwanted interference may consider the appropriate shielding of the sensor [23]. Another drawback of amplitude sensing is its vulnerability against significant temperature variations; the notch depth depends on resistance properties, which are influenced by temperature through the temperature coefficient of resistance (TCR).

Another important feature of the proposed sensing approach for spatial variables is that measurement can be contactless (by means of an air gap between the transmission line and the resonator). Since mechanical friction is avoided, contacless sensors benefit from aging effects less aggressive than in other sensors where there is friction, such as in rotary potentiometers [18].

The aforementioned sensing mechanism was proposed for the first time in [24] (Sect. 6.3) applying displacements to the resonator. Soon after, other sensor designs were reported in the literature [25–27].[8] Sections 6.3–6.6 are devoted to the reported spatial sensors for different spatial variables. The proposed spatial sensors are designed within the range 1–4 GHz (i.e. in the low regime of microwaves). A reason is because of the available in-house facilities for fabrication (a milling machine and a standard photo mask/etching process with a minimum strip and slot width of 150–200 μm). Another reason concerns the higher availability of inexpensive microwave components with a view to designing the feeding and readout circuits. Nonetheless, scaling the dimensions down is possible if necessary (see Fig. 6.4) using an appropriate fabrication technology.

6.2.2 Resonance Frequency Splitting/Shifting

The proposed sensing concept relies on transmission lines loaded with pairs of identical resonators whose principle of operation was described in Sect. 4.4. As mentioned therein, the structure functionality is based on resonance frequency splitting/shifting. Thus, in the reference (symmetric) state, the structure exhibits a single notch, but two notches appear when symmetry is disrupted, and the frequency separation between them is related to the level of asymmetry.

Dielectric loading may be a strategy to rupture symmetry. One possibility is to add a material or substance in a specific test region in the vicinity of the resonators. If the inclusion exhibits a symmetric distribution of the permittivity, there is a single transmission zero. On the contrary, if the dielectric loading is not symmetric, this can be detected by the appearance of two notches in the transmission coefficient.[9] Another

[8] In these designs, the main motivation is to obtain a frequency-independent notch with the displacement that enables the sensors to operate at a single fixed frequency. Otherwise, a frequency sweeping is necessary (for notch frequency tracking), which increases the complexity of the feeding and readout systems.

[9] For the sake of simplicity, the working principle is explained by symmetric/asymmetric distributions. To be rigorous, the principle should be actually described in terms of the absolute perturbations applied to each resonator. This means that a single notch can be still present for asymmetric inclusions that affect equitably the two resonators.

interesting possibility, which is the one considered, is to load each resonator with a dielectric load. In this case, no difference exists between the two dielectric samples if a single transmission zero is monitored. By contrast, the appearance of two transmission zeros in the frequency response indicates that different dielectric loads are introduced. By that means sensors and comparators able to perform real-time measurements of differential permittivity can be envisaged. A *differential sensor* should deliver the difference between two permittivity values. Conversely, a *comparator* (differential by definition) simply compares two permittivities indicating whether their values are the same or not. A differential sensor of permittivity may be useful for the analysis or characterization of the permittivity of materials and substances. A permittivity comparator may be interesting for the detection of defects or alterations in material samples by comparison to a well-known reference. Therefore, the proposed approach can be an alternative to sensors for organic tissues or microfluidic channels that have been proposed recently [7–13].

Clearly, the electrical sensing variable in this approach based on resonance frequency splitting/shifting is frequency.[10] When a resonance frequency is inspected, it should be noted that a rounded transmission notch introduces uncertainty in the measurement, so that a sharp transmission notch is more convenient. It should be highlighted that even though the measurand is sensed by inspection of resonance frequencies, the proposed sensing approach performs a differential measurement. The advantage of differential measurements is their ability to reject random errors (in this case due to frequency shifts) caused by changing ambient factors (that are applied to the whole structure). Hence, in differential measurements, the cross-sensitivities to ambient factors exhibit some degree of immunity. In fact, differential techniques are widely used to improve the performance of sensors [16–18]. Therefore, in the frequency-splitting/shifting approach, high robustness is expected by virtue of a differential frequency sensing.

Section 6.7 presents the proposed differential sensor and comparator for permittivity measurements [28]. To the best knowledge of the author, this is the first sensor consisting of transmission lines loaded with pairs of resonators whose symmetry is broken by means of inclusions. The first reported transmission line loaded with pairs of resonators, whose symmetry is disrupted for sensing purposes, was used as a displacement sensor by Horestani et al. [22].

6.3 Linear Alignment and Displacement

In this section we show how one-dimensional (1D) linear (along a straight line) alignment and displacement sensors can be designed on the basis of the symmetry properties described in Sect. 4.3. The displacement is sensed only in terms of magnitude (the sensing of the direction for linear displacements is addressed in Sect. 6.4).

[10]Besides the frequency, the depth of the resonance can also be a sensing magnitude.

6.3.1 SRR- and FSIR-Loaded Coplanar Waveguides

The response of a coplanar waveguide (CPW) loaded with a split-ring resonator (SRR), a folded stepped-impedance resonator (FSIR), or a single split-ring resonator was analyzed in Sect. 4.3.1 [24, 29, 30]. As concluded, if these resonators are etched symmetrically on the back side of the line, both the magnetic and the electric field components illuminating the resonators exactly cancel (there is no net electric or magnetic field flux over the resonator surface), and the particles are not driven at the fundamental resonance. This situation does not hold if symmetry is broken and, therefore, a possible lack of alignment and a relative linear displacement between the line and the resonator can be detected and measured, respectively. In the reported equivalent circuit model (Fig. 4.33), where the coupling modeled is the electric one only, it is clearly shown that such a coupling is sensitive to the displacement.

Reference [24] presents the first proof-of-concept sensor based on the above-mentioned symmetry properties, which is shown in Fig. 6.1. The sensing is performed by inspection of the notch magnitude of the transmission coefficient, S_{21}. Therefore, the sensitivity is defined as the variation of the notch magnitude with the displacement, i.e.

$$S = \lim_{\Delta y \to 0} \frac{\Delta 20 \log_{10} |S_{21}(f_0)|}{\Delta y} (\text{dB/m}), \tag{6.1}$$

where Δy denotes the linear lateral (along the transverse direction) displacement, and f_0 is the transmission zero frequency. Obviously, it is important to obtain a high sensitivity of the notch magnitude with the variable that destroys the symmetry of the structure, in this case a displacement. The key point in [24] is to validate the proof-of-concept for this novel sensing principle. Thus, an SRR is printed on the back side of the substrate rather than considering a sensing device with a movable SRR, which

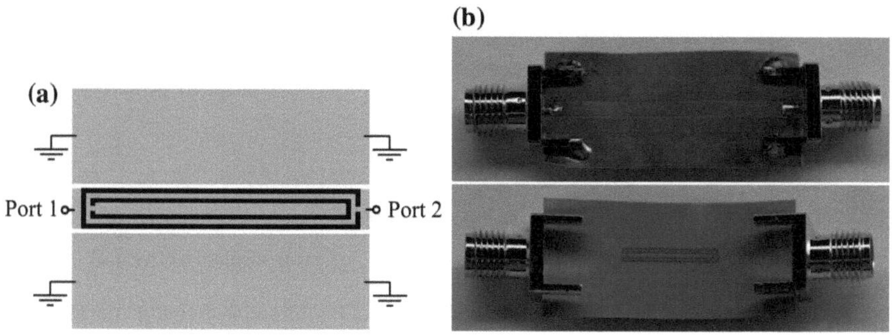

Fig. 6.1 **a** Layout and **b** *top* and *bottom* photographs of a CPW symmetrically loaded with a rectangular-shaped SRR. In the fabricated structure, access transmission-line sections are introduced. Dimensions are (see Fig. 4.6): for the CPW, $l = 3$ cm (fabrication), $W = 1.67$ mm and $G = 0.2$ mm (50-Ω line); for the SRR, $c = d = 0.2$ mm, $l_1 = 12$ mm, and $l_2 = W = 1.67$ mm. The substrate is *Rogers RO3010* with thickness $h = 127$ μm, relative permittivity $\epsilon_r = 10.2$, and loss tangent $\tan\delta = 0.0023$

is technologically more complex. As a result, the sensitivity is dependent on the line and the resonator geometries, but also on the substrate thickness. In order to achieve a high sensitivity, it is necessary to use very thin substrates (a 127-μm thickness is used) to enhance the line-to-resonator coupling (in a practical sensor this means that the resonator and the line should be as close as possible). Thereby stronger notches are obtained for a particular displacement. With regard to the geometry, the transverse dimension of the resonators should not excessively extend beyond the slot regions of the CPW. Otherwise, neither the electric nor magnetic fields illuminating the SRR exactly cancel if symmetry is broken, but the effects of asymmetry are not significant. Thus, a rectangular-shaped SRR is etched just below the central strip of the CPW (the resonance frequency may be adjusted with the longitudinal length).

Several fixed SRR-loaded CPWs with different lateral displacements are fabricated (see Fig. 6.1 for the aligned structure). Figure 6.2 shows the transfer curve of the sensor, where the characterization for the second resonance frequency is also included. For the second resonance, the sensitivity is not as good as the one corresponding to the first resonance, the reason being that the ring producing the second resonance is smaller and further displacement is required to be affected by the effects of asymmetry (i.e. the inner ring needs more displacement to be beneath one of the slot areas of the CPW). In addition, Fig. 6.3 plots the resulting sensitivity. The maximum input dynamic range is dictated by the strip and slot widths of the CPW. Concerning our specific structure, the resulting dynamic range is about 0.6 mm (±0.6 mm if the two possible directions are considered), the longest transverse dimension of the sensor (taken as the distance between the CPW ground planes) being 1.87 mm. Within the dynamic range, the larger the lateral shift, the deeper the notch in the transmission coefficient, and the resulting linearity is moderate.

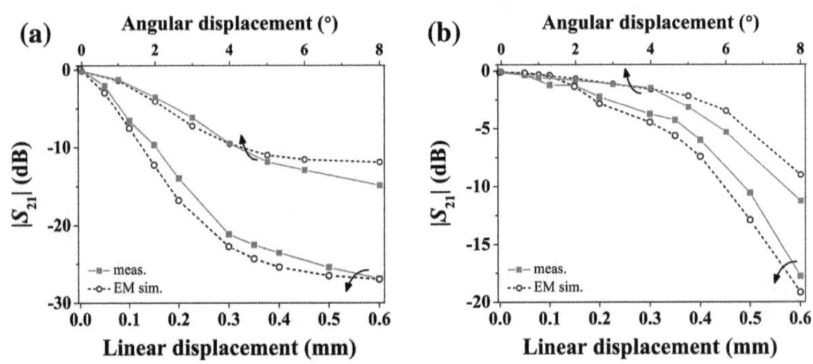

Fig. 6.2 Insertion loss at the **a** first and **b** second resonance for the proof-of-concept device shown in Fig. 6.1, resulting by either displacing laterally or rotating the SRR from its center. The notch frequencies are $f_0 \approx 2.3$ GHz and $f_0 \approx 2.7$ GHz for the first and second resonance, respectively. Reprinted with permission from [24]

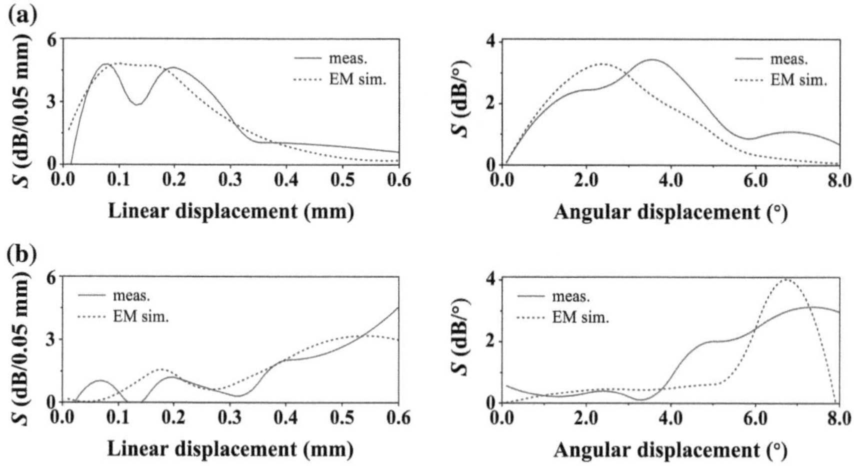

Fig. 6.3 Sensitivity at the **a** first and **b** second resonance computed from the transfer curves in Fig. 6.2. The sensitivity is calculated from discrete derivatives followed by a smoothing algorithm. The sign of the sensitivity is taken as positive when the displacement increases. Reprinted with permission from [24]

Fig. 6.4 Comparison between the notch depth given by lossy electromagnetic simulation for the structure of Fig. 6.1 and for the same structure with all dimensions scaled down. Reprinted with permission from [24]

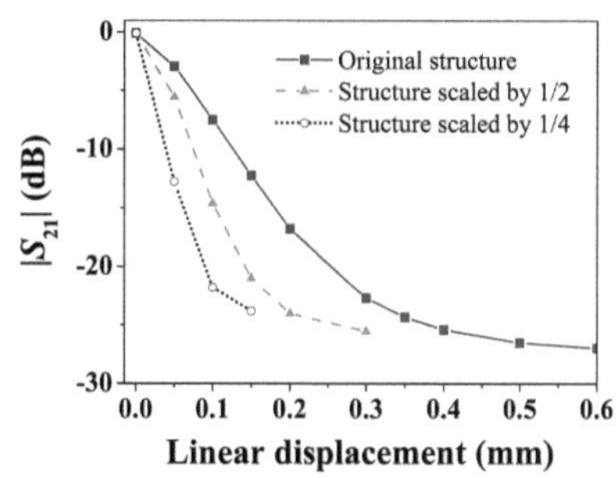

It is important to stress the possibility of sensitivity scaling, that is, by decreasing the overall dimensions of the SRR-loaded CPW by a particular factor, the sensitivity enhances almost proportionally to the scaling factor, as illustrated in Fig. 6.4. The reason is that, if we scale down the structure, the frequency response experiences a shift upward roughly maintaining the same notch magnitude (evidently regardless of the displacement). Such magnitudes would be identical (infinite) in the absence of losses. However, as the scaling factor increases, the notch depth is degraded by losses because losses increases with frequency [31]. Nonetheless, the limits on the scaling are dictated by the available fabrication technology.

As an alternative to the SRR, a proof-of-concept demonstrator based on a CPW loaded with an FSIR is presented in [30]. Again, a symmetric structure and additional structures with the FSIR laterally displaced are characterized. The sensor and its transfer curve can be seen in Fig. 6.5, proving the concept. The same substrate and dimensions as in the previous SRR-based sensor are considered for comparison

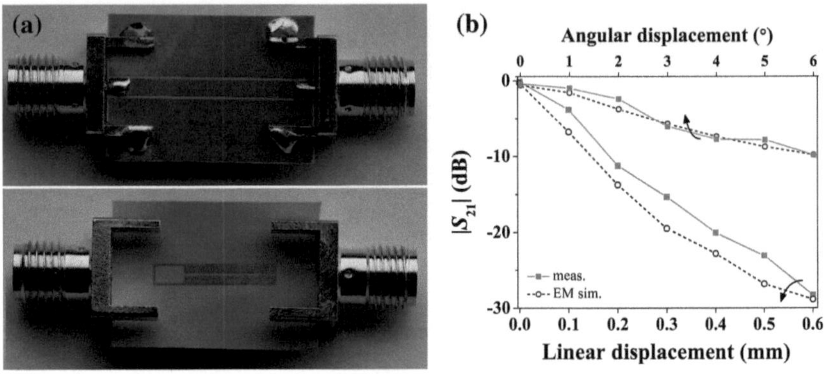

Fig. 6.5 **a** Fabricated CPW symmetrically loaded with an FSIR on the back side, and **b** notch magnitude with the FSIR laterally displaced or rotated from its center. The notch frequency is $f_0 \approx 1.25$ GHz. The substrate is *Rogers RO3010* with $h = 0.127$ mm, $\epsilon_r = 10.2$, and $\tan\delta = 0.0023$. Dimensions are: for the line, $l = 2$ cm, $W = 1.66$ mm, and $G = 0.2$ mm (50-Ω line); for the FSIR (see Fig. 2.23), $w_1 = 9$ mm, $w_2 = 0.2$ mm, $S = 0.2$ mm, $l_1 = 12$ mm, and $l_2 = 2$ mm. Reprinted with permission from [30]

purposes.[11] For a displacement of 0.3 mm, the measured notch magnitude is −21.1 and −15.3 dB for the SRR- and FSIR-based devices, respectively. However, a lower operating frequency and a better linearity within the dynamic range is obtained in the FSIR-based sensor.

6.3.2 SIR-Loaded Microstrip Lines

A description of the behavior of a microstrip line loaded with an SIR on an additional third layer was reported in Sect. 4.3.2 [32]. According to the proposed equivalent circuit model, the electric coupling between the line and the resonator plays a key role to strengthening the notch (depth and bandwidth) and to tuning the notch frequency as well. The sensitivities in magnitude and frequency can be expressed, respectively, as

$$S = \lim_{\Delta y \to 0} \frac{\Delta 20\log_{10}|S_{21}(f_0)|}{\Delta y} (\text{dB/m}), \tag{6.2a}$$

$$S = \lim_{\Delta y \to 0} \frac{\Delta f_0}{\Delta y} (\text{Hz/m}). \tag{6.2b}$$

where Δy denotes the displacement across the transmission line.

[11]The length of the SRR- and FSIR-based CPWs are different, 3 and 2 cm, respectively. The contribution of the slot mode generation to the frequency response is therefore not comparable in the two structures. In fact, in the FSIR-based CPW the length is reduced to mitigate the generation of the slot mode. To further decrease the slot mode generation, vias and backside strips could have been used.

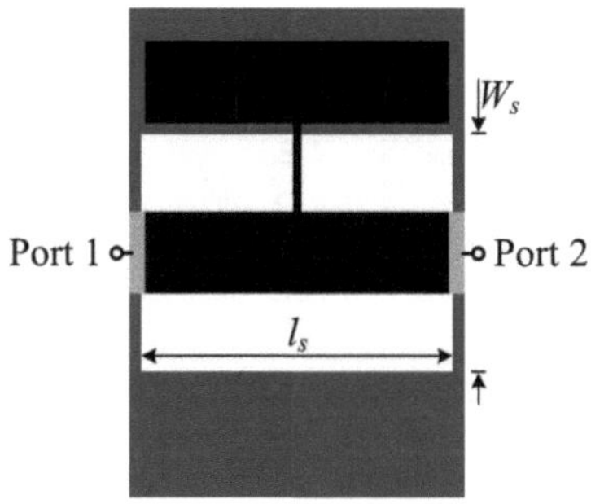

Fig. 6.6 Proof-of-concept demonstrator of an SIR-loaded microstrip line useful as linear alignment/displacement sensor (for a displacement of 2 mm corresponding to the strongest notch). The dimensions are: for the microstrip line, the strip width is $W = 2$ mm, $W_s = 5.8$ mm, and $l_s = 7.8$ mm; for the SIR, the wide and narrow widths are $w_1 = 7.6$ mm and $w_2 = 0.2$ mm, and the length of the sections are 2 mm. The top and bottom substrates are (see Fig. 4.35) *Rogers RO3010* with thickness $h = 0.254$ mm, relative permittivity $\epsilon_r = 11.2$, and loss tangent $\tan\delta = 0.0023$

With a view to designing alignment and displacement sensors by inspection of the notch depth and frequency, the line-to-resonator electric coupling should be as high as possible. In order to boost this coupling it is important to enhance the patch capacitance of the SIR to the line (as well as to the ground). It is for this reason that the microstrip line width is widened. Since this reduces the line characteristic impedance, in order to match the line to 50-Ω ports, a slot window in the ground plane underneath the signal strip is etched. This enhances the line inductance and drops the line capacitance simultaneously, compensating the reduction of the characteristic impedance caused by widening the strip. The slot is designed so that good impedance matching is obtained in the structure loaded symmetrically.

Following the previous design rules, Fig. 6.6 depicts the topology of a sensor prototype based on SIR-loaded microstrip lines designed and presented in [32]. The structure consists of a microstrip line structure stacked on bottom of another substrate where the SIR is etched on the top face (see the layer-cross section in Fig. 4.35). It is assumed that there is no air layer between the substrates to enhance the line-to-resonator electric coupling, but the lack of air prevents from relative motion of the SIR in regard to the line. As compared to the topology in Fig. 4.35b, the SIR sections are shortened. Thereby the transverse dimension is decreased, improving the sensitivity (this is at the expense of a higher resonance frequency and hence a larger electrical size through the path connecting the ports).

The prototype is characterized by several lateral displacements, as can be seen in Fig. 6.7. As is expected, the higher the displacement, the wider and the deeper the notch, and the lower the resonance frequency. The input dynamic range, that is equivalent to the displacement for maximum notch magnitude, is 2 mm (the longest transverse dimension of the sensor is that of the SIR which is 6 mm). The output dynamic range is 38.2 dB in amplitude and about 860 MHz in frequency (from 2.10 to 2.97 GHz). The results show that both the notch depth and the notch frequency are

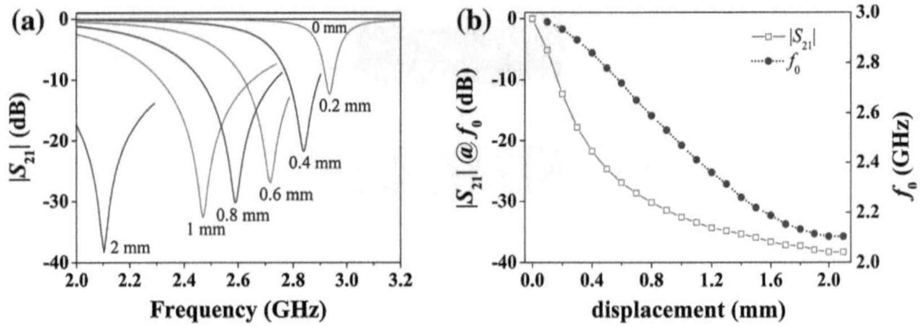

Fig. 6.7 **a** Transmission coefficient and **b** dependence of the notch magnitude and frequency with lateral displacement for the structure in Fig. 6.6, inferred from lossy electromagnetic simulation. Reprinted with permission from [32]

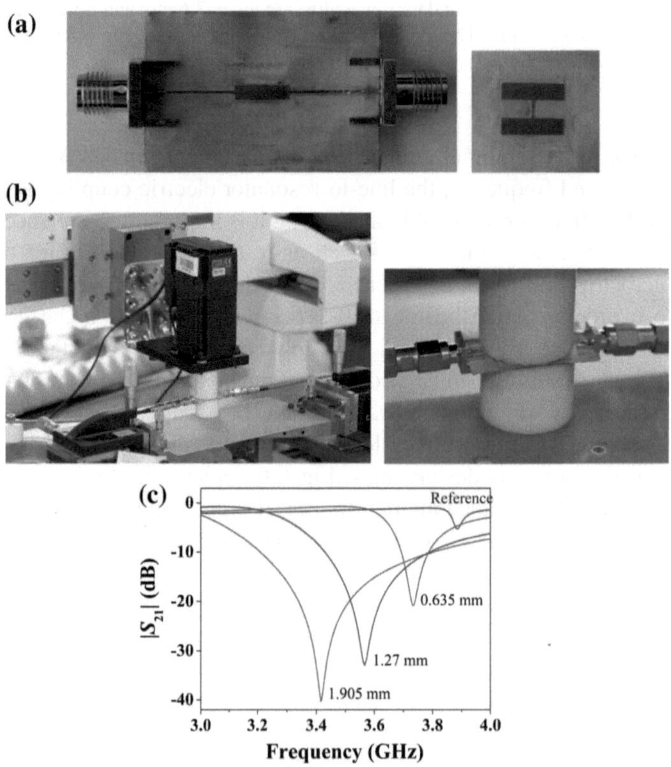

Fig. 6.8 **a** Photograph of the microstrip line (*top face*) and SIR (*top face*) shown in Fig. 6.6 where the line and resonator are printed on separate substrates, and where input/output lines of width $W = 0.2$ mm (50-Ω lines when only the bottom substrate is taken into account) are inserted to solder the connectors. **b** Photograph of the experimental set-up with the substrates attached to Teflon slabs (a material with a low relative permittivity, $\epsilon_r = 2.08$) and with manual positioners labeled in inches. For each relative displacement, the top substrate is separated from the bottom substrate, it is displaced, and it is stacked again onto the bottom substrate (the step motor is part of the experimental setup but is not used). **c** Measured transmission coefficient for some lateral displacements of the SIR. Reprinted with permission from [32]

very sensitive to the displacements, which is indicative of a significant increase in the line-to-SIR-to-ground capacitive coupling with the displacement from the aligned position. The resulting sensitivity is particularly very high in notch magnitude for small displacements, and this makes the sensor very suitable for alignment purposes. Specifically, the average sensitivity in notch magnitude is 5.4 dB/100 μm in the range up to 400 μm. The average sensitivity in notch frequency is 43 MHz/100 μm in the full dynamic range.

The previous proof-of-concept demonstrator is fabricated and measured in an experimental set-up that enables three-dimensional (3D) calibrations through manual positioners (see Fig. 6.8). It should be highlighted that it was not possible to perfectly align the SIR with the line. Instead, the displacements are respect to the sample referred to as reference. The problem of aligning manually (even with positioners) the SIR with the line is due to the fact that the sensor features a very high sensitivity for small asymmetries, and this points out its remarkable resolution for alignment capabilities. Another deviation observed in the experimental data is that a considerable upward frequency shift is obtained, and it is attributed to a very thin but non-negligible air layer between the substrates (which invariably drops the equivalent capacitance C_s in Fig. 4.36). Nevertheless, the sensing concept is experimentally validated.

6.4 Linear Position

The sensors presented in the preceding section used to detect the lack of linear alignment and to measure linear displacements apply to a one-dimensional (1D) space. Furthermore, those sensors are capable of sensing only the displacement magnitude, but not the direction of motion.[12] In this section it is shown that the sensing capability can be easily extended to a two-dimensional (2D) space, and the direction of displacement is also measurable. As a result, a 2D alignment and position sensor for linear displacements is presented.

6.4.1 *Coplanar Waveguides Loaded with Split Rings*

Reference [29] may be regarded as an expanded work of [24], where the sensing capability is extended to a 2D space and the direction of displacement is also detected. With a view to extending the sensing capability to a 2D space, a folded CPW (with a right-angle bend) loaded with two SRRs (actually single split rings)[13] are used, one

---

[12] As proposed in [6], the direction can actually be measured, but the dynamic range of the physical spatial quantity drops to a half. Additionally, the horizontal axis origin of the transfer curve does not correspond to the case with transparent propagation, that is, a transmission notch is present.

[13] Exclusively, the acronym SRR in this subsection refers also to a single split ring resonator.

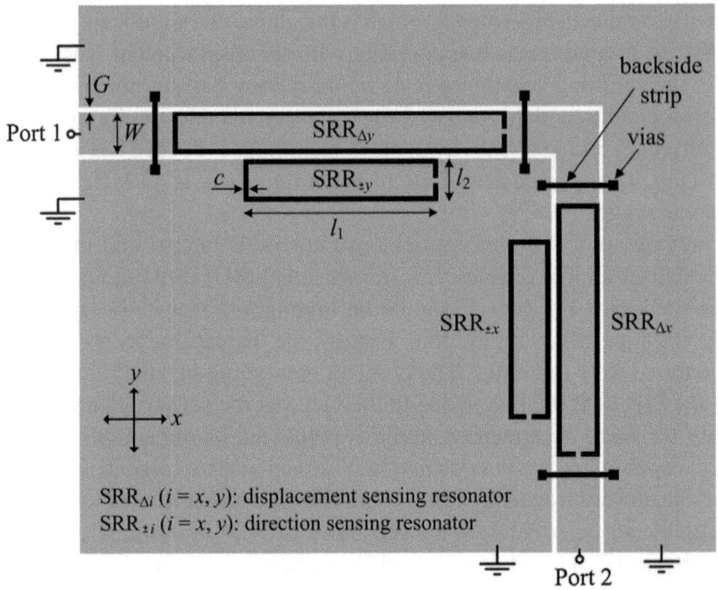

Fig. 6.9 Layout of the proposed proof-of-concept sensing device for the aligned position (i.e. the CPW and the displacement sensing SRRs are aligned). The substrate is *Rogers RO3010* with $h = 0.127$ mm, $\epsilon_r = 10.2$, and $\tan\delta = 0.0023$. Line dimensions are: $W = 1.67$ mm and $G = 0.2$ mm (50-Ω line). Resonator dimensions are: $c = 0.2$ mm, $l_1(\text{SRR}_{\Delta x}) = 9.95$ mm, $l_1(\text{SRR}_{\pm x}) = 7.05$ mm, $l_1(\text{SRR}_{\Delta y}) = 13.4$ mm, $l_1(\text{SRR}_{\pm y}) = 7.8$ mm, and $l_2 = W$

in each orthogonal direction (see Fig. 6.9). These SRRs, referred to as *displacement sensing resonators*, must be tuned at different frequencies to discriminate the two spatial dimensions in the frequency response. In order to provide the direction of motion, two additional resonators referred to as *direction sensing resonators* (also resonating at different frequencies) are utilized, one for each spatial dimension. However, for the aligned configuration, these resonators are etched beneath one of the CPW ground planes, near a CPW slot.[14] By that means, if the displacement direction drives such additional resonators toward the slot of the CPW, this is detected by a notch at the resonance frequency of these resonators. Conversely, by shifting these SRRs to the opposite direction such notches do not appear. This is the principle for sensing the direction of linear displacements.

For a better comprehension of the principle of operation of the proposed sensor, the four different displacements (rightward, leftward, upward, and downward) from the aligned structure are schematically illustrated in Fig. 6.10. The resonance frequencies of the four SRRs are denoted by $f_{\Delta y}$, $f_{\Delta x}$, $f_{\pm y}$, and $f_{\pm x}$. It can be seen

[14]The direction sensing resonators are placed on one of the sides of the displacement sensing resonators toward miniaturization. In the event that coupling between these two resonators affects sensor performance, a solution is simply to load them on different and sufficiently distant longitudinal locations of the CPW, evidently increasing the sensor size.

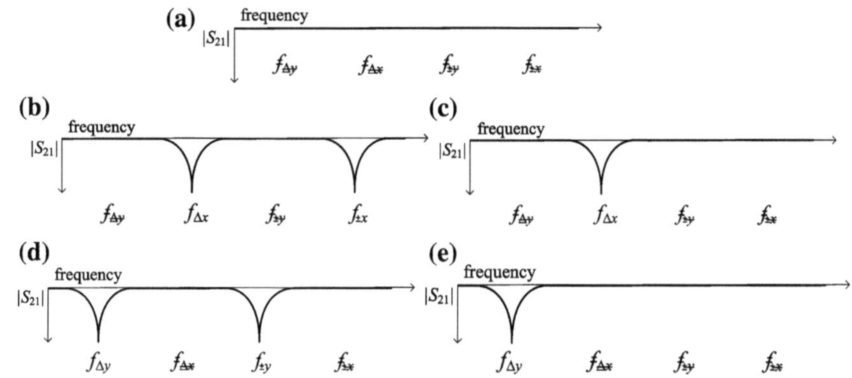

Fig. 6.10 Schemes indicating the primitive shifting operations and the resulting transmission coefficient magnitude, $|S_{21}|$. A notch is indicative of an SRR excitation. A linear displacement in the $\pm x$- and $\pm y$-orientation is indicated as $\pm\Delta x$ and $\pm\Delta y$, respectively, relative to the aligned position ($\Delta x = \Delta y = 0$). **a** $\Delta x = \Delta y = 0$. **b** $+\Delta x$ [$\Delta y = 0$]. **c** $-\Delta x$ [$\Delta y = 0$]. **d** $+\Delta y$ [$\Delta x = 0$]. **e** $-\Delta y$ [$\Delta x = 0$]

Fig. 6.11 Transmission coefficient of the proof-of-concept sensor for $\Delta x = +0.3$ mm and $\Delta y = +0.25$ mm. Reprinted with permission from [29]

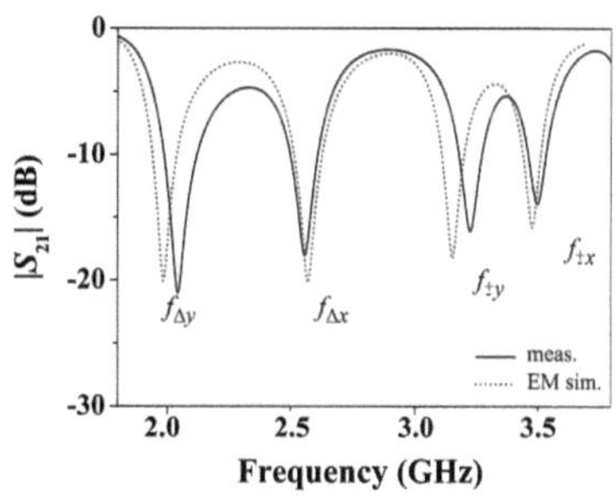

that displacements in the $\pm x$- and $\pm y$-direction can be measured in magnitude (by $f_{\Delta x}$ and $f_{\Delta y}$) and direction (by $f_{\pm x}$ and $f_{\pm y}$). Any other linear displacement is a combination of the previous ones, and hence it can be sensed as well. As an illustration, Fig. 6.11 shows the transmission coefficient of the proposed sensing structure for displacements of $\Delta x = +0.3$ mm and $\Delta y = +0.25$ mm. Note that four frequencies must be monitored within a bandwidth that ranges approximately from 1.8 to 3.8 GHz. The sensitivities are simply given by

$$S_x = \lim_{\Delta x \to 0} \frac{\Delta 20\log_{10}|S_{21}(f_{\Delta x})|}{\Delta x} (\text{dB/m}), \tag{6.3a}$$

$$S_y = \lim_{\Delta y \to 0} \frac{\Delta 20\log_{10}|S_{21}(f_{\Delta y})|}{\Delta y} (\text{dB/m}), \tag{6.3b}$$

where S_x and S_y denote the sensitivities in the x- and y-oriented directions, respectively.

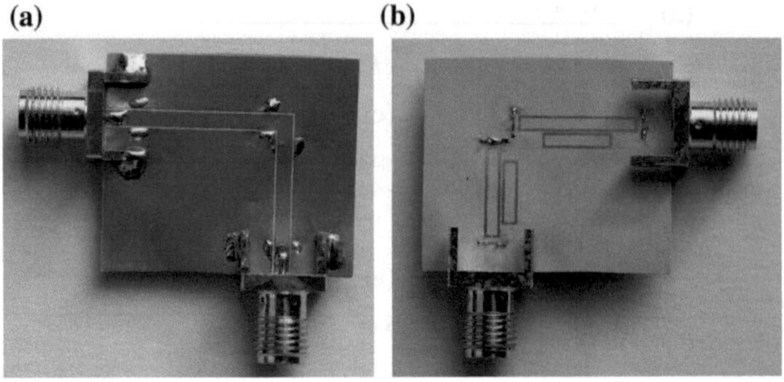

Fig. 6.12 Photograph of the proposed sensing device for the aligned position. **a** Top and **b** bottom views

The considered substrate and CPW (strip and slot widths) parameters are the same as in [24]. However, in this 2D sensor, split-ring resonators composed of only one ring are considered. This is due to two reasons; (*i*) because for thin substrates (as the one used) the coupling between the inner and the outer ring of SRRs is negligible (Sect. 4.3.1); and (*ii*) because the second resonance frequency of a single split ring is located beyond the one of SRRs. The latter feature is important to avoid (or reduce) interference between the four transmission notches corresponding to the four different resonators.

The proposed sensing structure is validated by considering several proof-of-concept structures with different displacements, where the resonators are printed on the back side of the substrate (see Fig. 6.12). Vias and backside strips are used to prevent the parasitic slot mode. Obviously, in a real sensor, the SRRs must be etched on a different substrate in order to achieve relative motion between the sensing SRRs and the bended CPW, but this complicates the measurement from a mechanical viewpoint. Nevertheless, the reported proof-of-concept measurements suffice to illustrate the potential of the approach. Positive and negative displacements in the x-direction (horizontal shift), as well as in the $x = y$-direction (diagonal shift), are characterized. The dependence of the notch magnitude (simulated and measured) for these displacements can be seen in Fig. 6.13. In reference to Fig. 6.13a, as expected, for positive displacements, the $SRR_{\pm x}$ is activated as is manifested by a clear increase in the notch magnitude at $f_{\pm x}$, whereas the specified −3-dB threshold level is not exceeded for negative displacements (indicating that the shift is in the negative direction). On the other hand, the notch corresponding to $f_{\Delta y}$ is approximately 0 dB, which indicates that the structure is aligned with the y-axis.[15] The input dynamic range is characterized up to ±0.3 mm, and for the maximum displacement the measured notch amplitude is approximately −19 dB. The dependence of the notch magnitude

[15]For $\Delta x = +0.3$ mm the measured notch is slightly below −3 dB because for this sample the y-axis position sensing resonators are somewhat misaligned due to the fabrication process (we experienced some difficulties to align the top CPW and the bottom SRRs in the in-house fabrication). On the other hand, this is an indicator of the high sensitivity of the approach to detect misalignments.

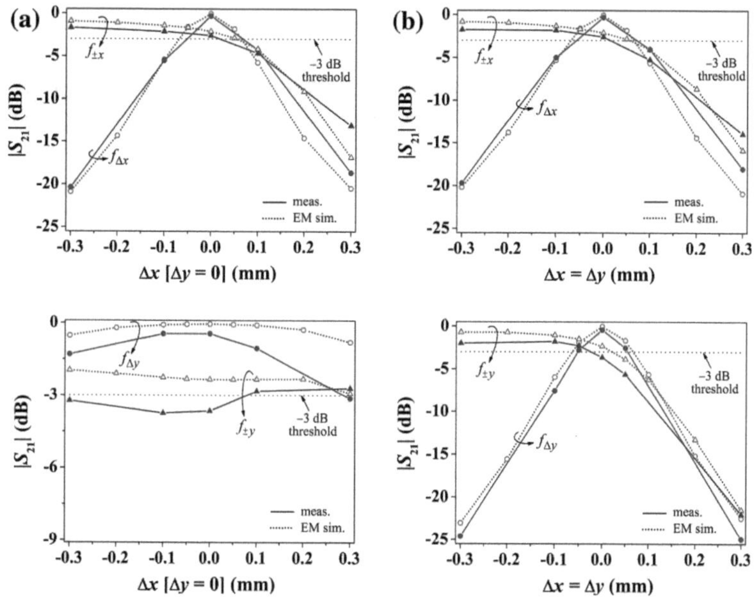

Fig. 6.13 Notch magnitude in the transmission coefficient at the indicated resonance frequencies **a** for x-oriented displacement and **b** $x = y$-oriented displacement. Results for (*top*) x- and (*bottom*) y-axis position sensing. Reprinted with permission from [29]

is similar and roughly linear in the two orthogonal directions. In the light of these results, based on a partial performance characterization that is representative of the potentiality and validity of the sensing proposal, the proposed 2D linear alignment and position sensing approach is verified.

It is finally important pointing out that the notch magnitude associated to a displacement sensing resonator depends not only on the displacement, but also on resonator dimensions and inter-notch interference.[16] This causes that, for the same displacement, the notch magnitude of the y-axis displacement sensing resonator is deeper than that of the x-axis.

6.5 Angular Alignment and Displacement

This section presents sensors to detect the lack of angular alignment and to measure the angular displacement (i.e. rotation) between two objects. To this end, the resonator axis is perpendicular to the line axis intersecting it, and different angular orientations of the resonator are considered. The proposed sensors are able to

[16]Inter-notch interference limits the performance (e.g. the input dynamic range) of the proof-of-concept sensor. Better performance is expected using folded SIRs (FSIRs) since further increase in the ratio between the second and first resonances can be achieved.

measure the displacement magnitude, but not the direction (without reducing the dynamic range of measurement by half [6]).

It is interesting to mention that microwave rotation sensors based on frequency [6] and phase shift [33], with good linearity and higher dynamic range (180° and 360°, respectively) than those proposed here, have been recently reported. However, these sensors cannot be extended to the implementation of angular velocity sensors with feeding and readout circuits as simple as with amplitude levels (next section).

6.5.1 SRR- and FSIR-Loaded Coplanar Waveguides

As demonstrated in Sect. 6.3.1, SRR- or FSIR-loaded CPWs can be useful for the implementation of linear alignment and displacement sensors. Additionally, these structures can also be designed to operate as angular alignment and displacement sensors. In [24, 30], the same SRR- and FSIR-loaded CPWs that are characterized with linear displacements, are also evaluated by applying small rotations (up to 6°). Several SRR- and FSIR-loaded CPWs with different angular displacements are fabricated to validate the approach (the resonators are etched on the back side of the substrate). Photograph of the symmetric structures are shown in Figs. 6.1b and 6.5a, while the transfer functions that relate the notch magnitude with the rotation are plotted in Figs. 6.2 and 6.5b. Figure 6.3 shows the sensitivity for the SRR-based rotation sensor. Since the dynamic range of angular displacements is very small (about 6°), the proposed sensors may be especially suitable for angular alignments. For a rotation of 6°, the measured notch magnitude is −12.7 dB (SRR-based sensor) and −9.6 dB (FSIR-based sensor).

6.5.2 ELC-Loaded Coplanar Waveguides

As explained in Sect. 4.3.3, in ELC-loaded CPWs the frequency response (in the vicinity of the fundamental resonance) depends on the angular orientation of the resonator. Specifically, due to the electric and magnetic walls of the ELC resonator, CPWs loaded with this resonator alternate total transmission or a transmission zero by rotating the resonators 90°. These transmission characteristics are interesting to implement angular alignment and displacement sensors by measuring the attenuation peak in the transmission coefficient.[17] The sensing principle is based on the fact that the line-to-resonator magnetic coupling level is sensitive to the angular displacement, as demonstrated by (4.10).

---

[17] The notch amplitude is chosen as the electrical variable to measure the angular displacement. Additionally, or alternatively, other electrical variables (e.g. phase or frequency) may be used, but the characterization with the amplitude allows for implementing the simple angular velocity sensor system that is presented in the next section.

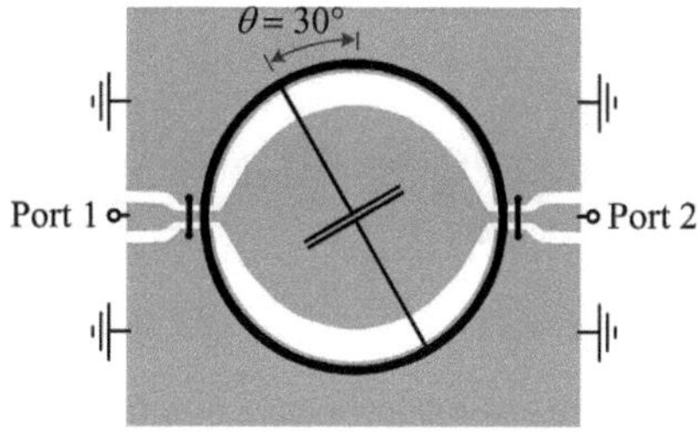

Fig. 6.14 Layout of the angular displacement sensor presented in [34] for $\theta = 30°$. The substrate is *Rogers RO3010* with $h = 1.27$ mm, $\epsilon_r = 11.2$, and $\tan\delta = 0.0023$. The dimensions are: for the line, W and G are tapered such that the characteristic impedance is 50 Ω (the radial distance from the center of the CPW to the lateral ground planes is 7.6 mm); for the ELC resonator (see Fig. 2.25): mean radius $r_0 = 8.05$ mm, $w_1 = 6$ mm, $l_1 = w_2 = s = 0.2$ mm, and $w_3 = 0.5$ mm. Access lines to solder the connectors are added resulting in a non-electrically small CPW. For such a reason and, because the CPW-based structure is asymmetric for most of the rotation angles, vias and backside strips at each side of the resonator are used to suppress the parasitic slot mode

An angular alignment and displacement sensor based on ELC-loaded CPWs is presented for the first time in [34], and is shown in Fig. 6.14. A circularly-shaped ELC resonator in combination with a circularly-tapered CPW is designed. By this means, as was shown in Sect. 4.3.3, good linearity in the notch magnitude with the rotation angle is obtained. The sensitivity is the variation of the notch magnitude with the rotation, i.e.

$$S = \frac{d20\log_{10}|S_{21}(f_0)|}{d\theta}(\text{dB}/°), \tag{6.4}$$

and is approximately constant. In order to maximize the output dynamic range, the notch depth should be increased. To this end and, according to the reported circuit model in Sect. 4.3.3, the ELC capacitor is designed small (so that the electrical size of the resonator increases).

In the first place, the sensing principle is validated with different resonator orientations by some conventional fixed resonator-loaded lines, where the resonator is etched on the back side of the substrate (see Fig. 6.15). However, for sensing purposes, the resonator must be attached to another substrate in order to allow for a relative motion between the resonator and the CPW. Then, in the second place, a more sophisticated prototype shown in Fig. 6.16 is developed where the CPW and the resonator are printed on different substrates in parallel planes with an in-between air layer.[18] The resonator substrate is suspended by another substrate attached on

[18]The smaller the distance between the CPW and the resonator, the higher their mutual magnetic coupling. However, for tiny distances, the increase in the dynamic range may be at the expense of some insertion loss for 0°, caused by a strong influence of the resonator to the line parameters (inductance and capacitance) that produces impedance mismatch at the 50-Ω port interfaces. The distance is set to 1.27 mm, providing a good balance between the insertion loss at 0° and the dynamic range.

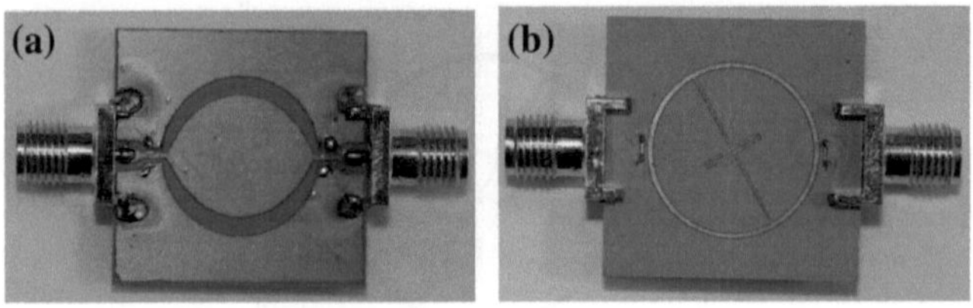

Fig. 6.15 Photograph of the **a** top and **b** bottom faces of a fabricated ELC-loaded CPW (for a specific rotation angle of $\theta = 30°$)

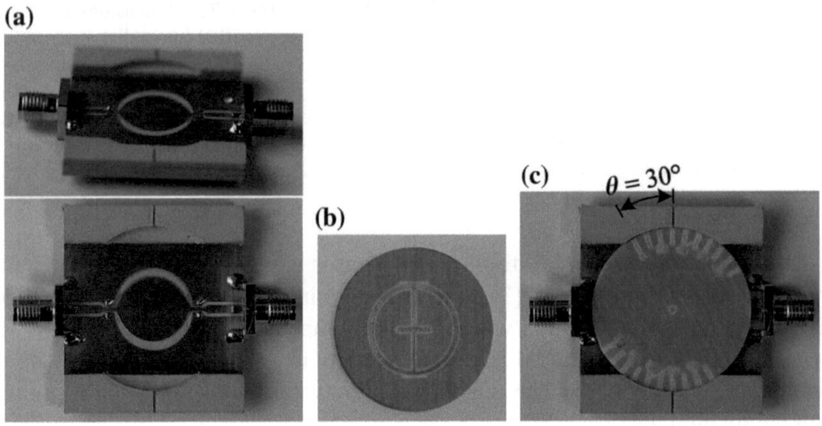

Fig. 6.16 Rotation sensor with the resonator etched on a suspended movable substrate, separated 1.27 mm from the CPW metal layer. **a** CPW structure. **b** ELC resonator. **c** Whole sensor implemented with a suspended movable substrate where an angular grid is used for angle identification. The movable substrate is *Rogers RO4003C* with $h = 0.8128$ mm, $\epsilon_r = 3.55$, and tan$\delta = 0.0021$

Fig. 6.17 Dependence of the notch magnitude with rotation for the structures of Figs. 6.15 and 6.16. The notch frequency is $f_0 \approx 1.21$ GHz (fixed resonator) and $f_0 \approx 2.29$ GHz (suspended resonator). Reprinted with permission from [34]

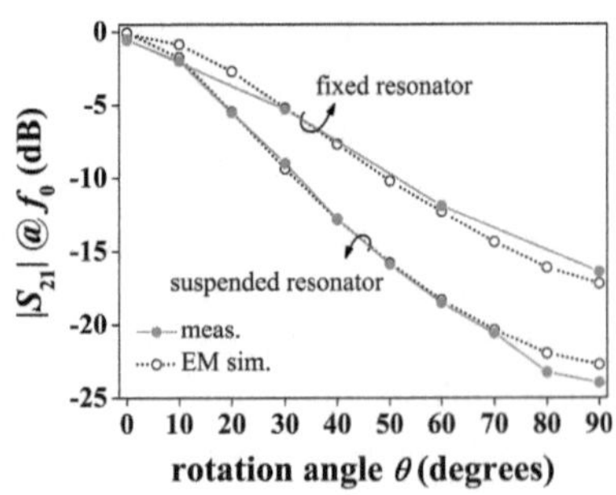

top of the CPW substrate, and is manually moved. The transfer curve of these two implementations (with fixed and suspended resonators) are plotted in Fig. 6.17.

Finally, in the third place, in [35], which is an expanded work of [34], the sensor is tested using a more realistic experimental set-up. The CPW and the ELC resonator are again etched on separate substrates, which can be seen in Fig. 6.18. A prototype sensor is characterized with the resonant element being adhered to a cylinder that is

Fig. 6.18 Photograph of the fabricated CPW and ELC resonator on separate substrates. Again, the substrates are: for the CPW, *Rogers RO3010* with $h = 1.27$ mm, $\epsilon_r = 11.2$, and $\tan\delta = 0.0023$; for the ELC resonator, *Rogers RO4003C* with $h = 0.8128$ mm, $\epsilon_r = 3.55$, and $\tan\delta = 0.0021$. The in-between air layer separating the CPW and resonator in the rotation measurement is set to 1.27 mm

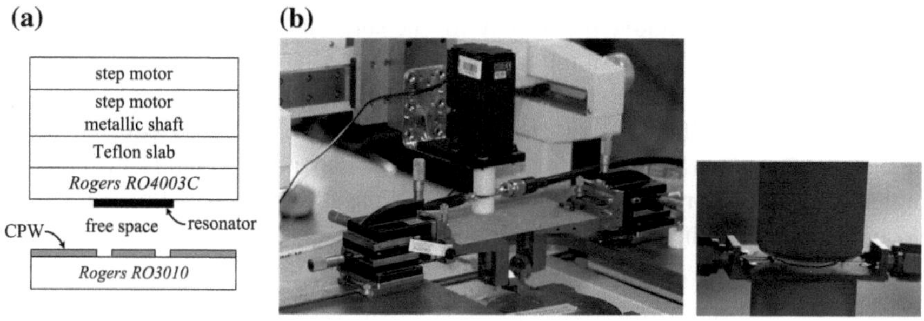

Fig. 6.19 **a** Layer cross-section and **b** photograph of the experimental set-up for the angular displacement measurement. First, a three-dimensional (3D) spatial calibration between the CPW and the resonator is carried out by means of manual positioners. Second, a step motor *STM 23Q-3AN* is used to accurately control the angular position. The parameters of Teflon are: $h = 3.5$ cm, $\epsilon_r = 2.08$, and $\tan\delta = 0.0004$. The air layer parameters are: $h = 1.27$ mm, $\epsilon_r = 1$, and $\tan\delta = 0$

rotated by a step motor, as shown in Fig. 6.19. In particular, the resonator substrate is attached to the metallic shaft of the step motor by means of a Teflon cylindrical slab.[19] The results of the characterization can be seen in Fig. 6.20. In comparison to the structure based on a suspended substrate, the notch frequency experiences a slight decrease of 30 MHz due to the Teflon slab, while the change in the notch depth is negligible. Interestingly, this feature indicates that the notch magnitude (4.12) possesses not only some robustness, but also less cross-sensitivity than the resonance frequency (4.1), to variations or tolerances on the permittivity. Certainly, the resonance frequency depends strongly on the resonator capacitance. The input dynamic range of measurable rotations is of course limited to 90° owing to the sensor symmetry, while the output dynamic range is 23.7 dB. The resulting measured linearity and sensitivity are 5.1% and 0.26 dB/° in average, respectively.

It should be pointed out that the resonance frequency as well as the notch magnitude experiences an increase when an in-between air layer is introduced between the CPW and the ELC resonator, as it must be in a real sensor system. The reason

[19]The Teflon slab is used to avoid close proximity between the resonator and the metallic shaft. Otherwise, the electrical performance of the sensor may change.

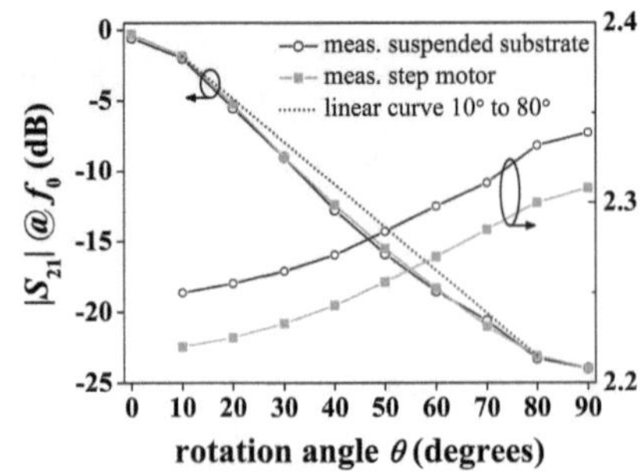

Fig. 6.20 Transfer function of the measured notch magnitude at the notch frequency versus the rotation angle for the structure in Fig. 6.18 characterized by the experimental set-up in Fig. 6.19. Reprinted with permission from [35]

is the decrease in the relative permittivity of the surrounding medium, which drops significantly the capacitance of the resonator.

Finally, it should also be stressed that the notch frequency is shifted from 2.22 GHz (10°) to 2.308 GHz (90°) due to the applied rotation, and thus frequency could be even used as a sensing variable. This frequency-dependent notch is attributed to angle-dependent resonator parameters (note that the structure cannot be considered electrically small at these frequencies).

6.5.3 ELC-Loaded Microstrip Lines

In Sect. 4.2.3 the electromagnetic behavior of a pair of microstip lines symmetrically loaded with an ELC resonator was analyzed. As explained therein, when the line pair is fed by an even-mode excitation, the structure exhibits either total transmission or a transmission zero by means of 90° particle rotation (see Fig. 6.21). Accordingly, rotations within a 90° dynamic range can be measured from a common-mode transmission coefficient, the notched response being directly related to the line-to-resonator magnetic coupling.

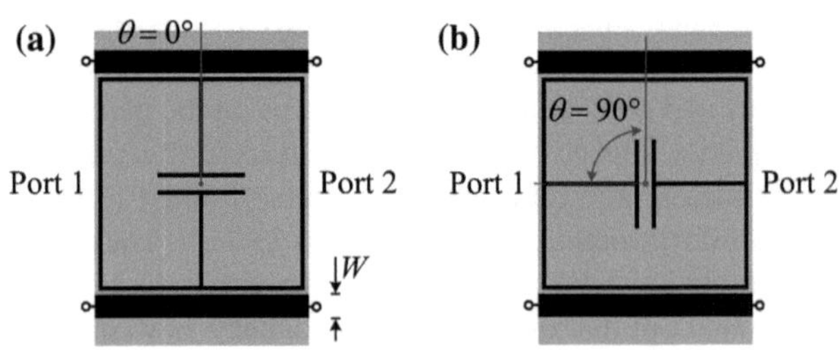

Fig. 6.21 Parallel microstrip lines loaded with an ELC resonator for the two extreme cases of angular orientation; the fundamental ELC resonance is **a** forbidden for $\theta = 0°$ and **b** allowed for $\theta = 90°$

(a)

(b)

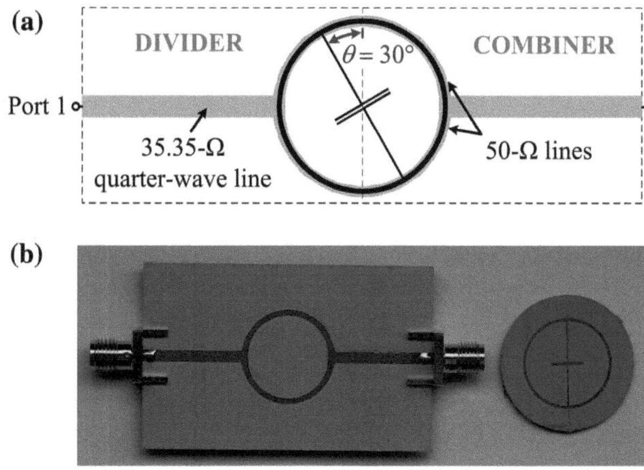

Fig. 6.22 **a** Layout and **b** photograph of the designed angular displacement sensor based on an ELC-loaded microstrip line. The substrates are *Rogers RO3010* with thicknesses $h = 1.27$ mm (microstrip line) and $h = 0.635$ mm (resonator), relative permittivity $\epsilon_r = 11.2$, and loss tangent $\tan\delta = 0.0023$. The line widths are 2.06 mm and 1.04 mm. With reference to Fig. 2.25, the ELC mean radius is $r_0 = 8.05$ mm, $w_1 = 6$ mm, $w_2 = l_1 = s = 0.2$ mm, and $w_3 = 0.5$ mm. The gap between the microstrip line and the resonator in the rotation measurement is set to 0.254 mm

Indeed, the behavior of this structure is basically the same as that of ELC-loaded CPWs as the ones used in the previous subsection [34, 35]. To clarify this, compare the circuit models in Figs. 4.41a and 4.23b for the case $\theta = 0°$, and those in Figs. 4.41b and 4.19b for $\theta = 90°$. Clearly, the circuits describing the microstrip- and CPW-based structures are formally identical.

An angular displacement sensor inspired by an ELC-loaded parallel microstrip line pair operating in common mode is designed in [36], and is shown in Fig. 6.22. However, rather than driving the pair of lines by two sources in an even-mode configuration, a T-junction power divider cascaded to another one acting as a combiner are used. By that means the structure is excited by a single and conventional single-ended microstrip line port; by feeding the structure with a single-ended signal, the divider acts as a single-ended to common-mode signal transition. Conversely, the combiner converts the common-mode signal into a single-ended signal.[20]

The resonator is the same circular one as in the preceding subsection, and the two dividers are accordingly circularly-shaped designed to improve linearity. The circular line sections exhibit a 50-Ω characteristic impedance.[21] The straight section of the combiner/divider is implemented with 35.35-Ω (Z_c) impedance inverters (quarter-wavelength lines at 1.45 GHz) to achieve impedance matching to 50-Ω (Z_0) ports, since the input impedance (at 1.45 GHz) seen into the straight section is

[20]This is absolutely true when the microstrip structure is unloaded. When the microstrip structure is asymmetrically loaded with the ELC resonator, mode conversion is generated, and the single-ended signal at the combiner output is a superposition of common- and differential-mode signals.

[21]The lines are uncoupled and the even- and odd-mode characteristic impedances are 50 Ω.

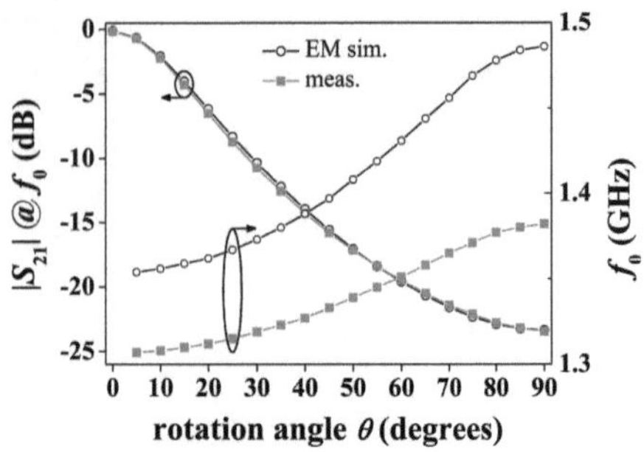

Fig. 6.23 Notch magnitude and frequency versus the angular displacement. The frequency shift is attributed to substrate- and fabrication-related tolerances (f_0 is very sensitive to these tolerances). Reprinted with permission from [36]

$$Z_{in} = \frac{Z_c^2}{Z_0/2} = 50\ \Omega. \tag{6.5}$$

The circular line pair is arranged face-to-face with the ELC external ring in order to enhance electromagnetic coupling.

Although undesired common-to-differential mode conversion may arise in the parallel microstrip line pair for asymmetric angular loadings, the differential mode is indeed short-circuited at the junctions of the dividers (i.e. at both sides of the resonator). However, if the circularly-shaped lines are electrically long,[22] non-negligible mode mixing may be generated along them. In order to reduce electrically the size of these lines at resonance, in contrast to the CPW-based sensor of the previous subsection, the resonance frequency of the resonator is decreased by etching it on a high relative permittivity substrate ($\epsilon_r = 11.2$).

The measurement set-up is that shown in Fig. 6.19 with the exception that the resonator is closer to the line (the air gap is 0.254 mm) to boost their coupling. The decrease in the line-to-resonator distance is because of the fact that the notch depth in this microstrip implementation is seen to be poorer than in the CPW implementation (for the same substrates and line-to-resonator distance). Figure 6.23 shows the transfer curve of the designed sensor. The measured output dynamic range is 23.3 dB, and the measured linearity and sensitivity are in average 9.7 % and 0.26 dB/°, respectively. As compared to the counterparts based on ELC-loaded CPWs, the microstrip implementation is an alternative that does not require air bridges, and the ground plane provides backside isolation.

6.6 Angular Velocity

The basic idea behind many velocity or acceleration sensors is a measurement of the displacement of an object with respect to some reference object. Therefore, displacement sensors are integral parts in many of them [18]. In this section it is

[22]The electrical path in the circular microstrip lines is longer than that for the circularly-tapered CPW of the previous subsection.

demonstrated that the angular velocity of a rotating object can be measured in a simple manner using, as a building block, the previous angular displacement sensors based on measuring amplitude levels. A novel angular velocity sensor concept based on symmetry properties, with high accuracy and precision, is proposed.

6.6.1 *ELC-Loaded Coplanar Waveguides and Microstrip Lines*

Assume that, in the ELC-based angular displacement sensors of the previous section, the resonator is rotating at a particular velocity. The angular velocity is the change of the angular position, θ, with time, t. In terms of the angular frequency, the velocity is

$$\omega_r = \frac{d\theta}{dt} = \frac{2\pi}{T_r} = 2\pi f_r, \tag{6.6}$$

where T_r is the rotation period and f_r is the corresponding linear frequency. Since the angular position varies with time, the notch magnitude as well. Therefore, the signal at the output of the displacement sensor is periodically amplitude modulated with time. Due to the displacement sensor symmetry, the transfer function is periodic every 180° varying between a maximum and a minimum attenuation (see the particular example in Fig. 6.24). As a result, by feeding the line with a harmonic signal at a frequency referred to as carrier, f_c, tuned in the vicinity of the fundamental resonance frequency,[23] the time distance between consecutive transmission peaks of the envelope signal, $T_m = 1/f_m = 2\pi/\omega_m$, corresponds to half a period of rotation. Therefore,

$$T_r = nT_m = 2T_m, \tag{6.7a}$$

$$f_r = f_m/n = f_m/2, \tag{6.7b}$$

$$\omega_r = \omega_m/n = \omega_m/2, \tag{6.7c}$$

where n is the number of transmission peaks per cycle, and the angular velocity can be rewritten as

$$\omega_r = \frac{\pi}{T_m} = \pi f_m = \omega_m/2. \tag{6.8}$$

In brief, the angular displacement sensor with a time-varying resonator angular orientation acts as an amplitude modulator, and the angular velocity can be inferred from the modulation using (6.8), that is, the transfer curve. As can be derived from (6.8), the sensitivity is $S = 2$, and hence is constant. The sampling rate of measurement is assumed to be discrete and given by T_m.

[23] Although the resonance frequency is sensitive to the displacement, for velocity sensing purposes, a single fixed frequency can also attain a transfer function (in notch magnitude) with a well-defined period necessary to measure the velocity.

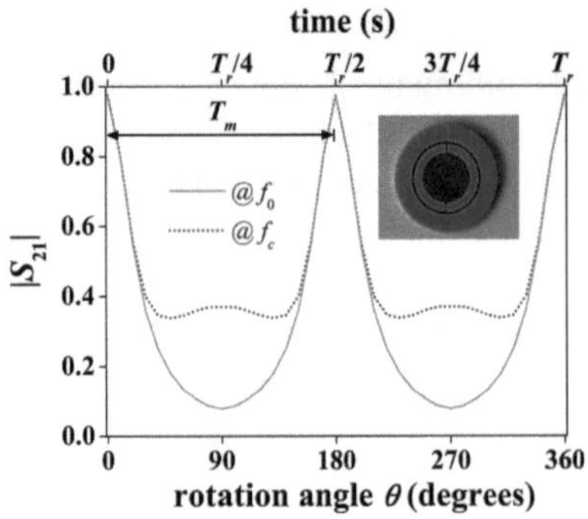

Fig. 6.24 Magnitude of the measured transmission coefficient in linear scale as a function of the angular position (or equivalently time if the resonator is rotating at a particular angular velocity) at the notch frequency, f_0, and at the carrier frequency, $f_c = f_0(\theta = 10°) = 1.515$ GHz, for the displacement sensor of Fig. 6.18 but with the ELC resonator of Fig. 4.42 shown in the inset. The experimental set-up is that in Fig. 6.19. Reprinted with permission from [35]

The proposed concept for angular velocity sensing is characterized and validated for the first time in [35]. To experimentally measure the angular velocity in time domain, a circulator (to avoid unwanted reflections) and an envelope detector (to obtain the envelope of the amplitude modulated signal) are cascaded to the output port of the angular displacement sensor. A schematic and an experimental implementation of the sensor system including the feeding and readout circuits is depicted in Fig. 6.25. The classical envelope detector circuit composed of a diode and a capacitor-resistor (RC) filter is used to demodulate the amplitude modulated signal. The details on the design and implementation of the detector can be found in Appendix F.

To prove the concept, the step motor is configured with a uniform angular velocity of $\omega_r = 2\pi f_r$, where $f_r = 1$ and 50 Hz (60 and 3,000 rpm, respectively, the latter being the maximum angular velocity of the motor). At the detector output, the envelope signal is visualized by an oscilloscope (see Fig. 6.26).[24] The measured linear velocities are $f_r = 0.998$ and 49.751 Hz in CPW implementation [35], and $f_r = 0.998$ and 50.251 Hz in microstrip technology [36]. These measured values are very close to the nominal velocities specified in the step motor (the total error is 0.2 and 0.5 % for 1 and 50 Hz, respectively), validating the sensing approach.

It is important to highlight certain features of the sensing proposal:

- The sensor is able to measure the speed (i.e. the magnitude) of the angular motion, but not the direction.
- Since for practical situation the carrier frequency is much higher than the linear frequency of rotation (or modulation), the measurable range of velocities is theoretically unlimited. Therefore, by virtue of a theoretical infinite input/output dynamic range, the transfer curve extends infinitely.

[24]Optionally, the envelope signal can be in addition converted into a digital rectangular signal (by an electronic comparator) to ease the signal processing of the readout system.

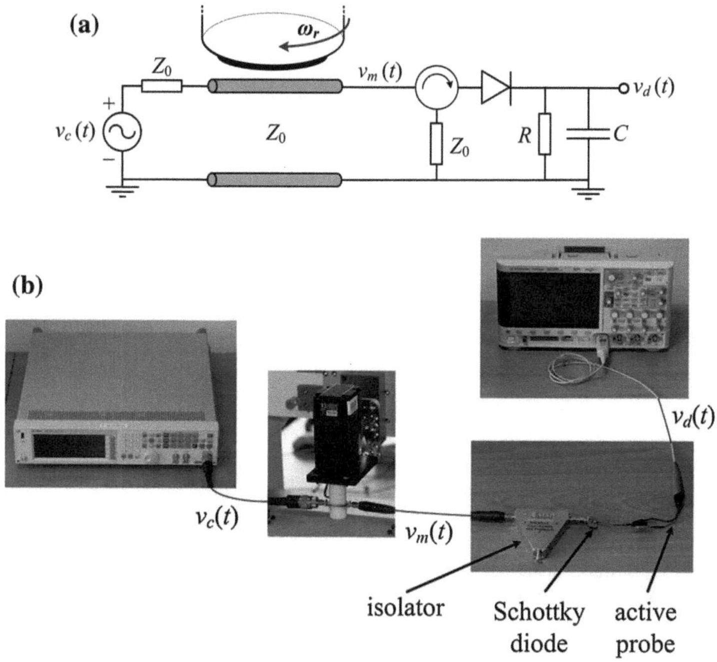

Fig. 6.25 Set-up for the angular velocity measurement. **a** Schematic and **b** photographs of the experimental implementation. The carrier signal is generated by an *Agilent N5182A* signal generator. The isolator is implemented with a broadband (1-2 GHz) coaxial *ATM ATc1-2* circulator, with one of the ports terminated with a matched $Z_0 = 50\text{-}\Omega$ load. The envelope detector is synthesized by a surface mount microwave Schottky detector diode *Avago Technologies HSMS 2860* and a single-ended active probe *Agilent N2795A* (with an input impedance of 1 MΩ in parallel with 1 pF). The envelope signal is recorded by an *Agilent 3054A* oscilloscope

- Since the sensitivity is constant, the envelope signal waveform is independent on the angular velocity (see Fig. 6.26).
- To enhance precision, the velocity measurement can be done between very distant non-consecutive peaks (or by averaging the time between adjacent peaks), although in this case the velocity must be constant.
- The signal generator, the circulator, the detector, and the oscilloscope can be replaced with low-cost circuits and components.
- Any structure capable of producing a periodically time-varying transmission coefficient with angular motion may be used. However, the advantage of the presented solution is that since an angular displacement sensor is used as a building block, an all-in-one sensor able to measure the displacement and the velocity may be possible.
- The resonator axis is required to be aligned as much as possible with the axis of rotation (the resonator must be etched on a plane parallel to the transmission line, e.g. at the top or bottom face of a cylinder).

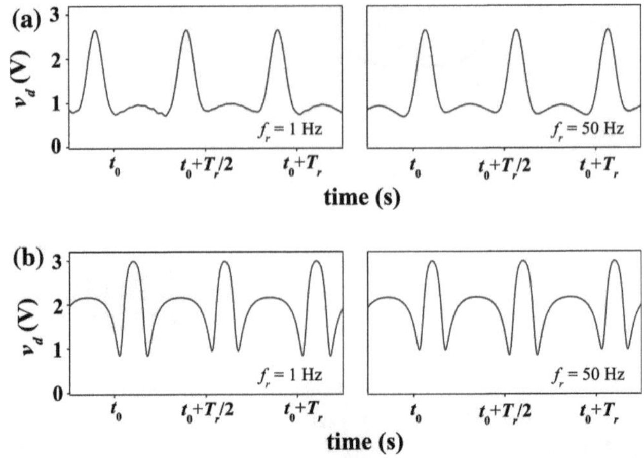

Fig. 6.26 Measured envelope signals using the experimental set-up in Fig. 6.25. **a** ELC-loaded CPW in Fig. 6.18 with the ELC resonator in the inset of Fig. 6.24 where $f_c = f_0(\theta = 10°) = 1.515$ GHz. **b** ELC-loaded microstrip line in Fig. 6.22 where $f_c = f_0(\theta = 10°) = 1.308$ GHz. The time t_0 is an arbitrary initial time. Ideally, $T_r = 1$ s ($f_r = 1$ Hz) and $T_r = 20$ ms ($f_r = 50$ Hz). Reprinted with permission from [35, 36]

Finally, it is important to mention that the proposed sensor system resembles (or is inspired by) an amplitude modulation (AM) system, in which a carrier signal (with frequency f_c) is amplitude modulated by an information signal of lower frequency (f_m) that is subsequently recovered through the envelope of the modulated signal (Appendix F).

6.7 Differential Permittivity

This section proposes a new concept of differential sensor and comparator for permittivity measurements whose principle of operation relies on the symmetry properties of transmission lines loaded with pairs of resonators (Sect. 4.4). Hence, differential measurements are possible in virtue of the resonance frequency splitting/shifting functionality.

6.7.1 Microstrip Lines and Conductor-Backed Coplanar Waveguides Loaded with Pairs of SISSs

The symmetry-based differential permittivity sensor proposed by Naqui et al. is reported in [28]. It consists of a microstrip line loaded with a pair of identical stepped-impedance shunt-stubs (SISSs) at the same junction, as was shown in Fig. 4.56a.

The behavior of this structure was already explained in Sect. 4.4.1, and in the present subsection the focus is on its capabilities for sensing differential permittivities.

By applying a capacitive perturbation to each SISS, it was demonstrated that no difference exists between the two perturbations if a single transmission zero is monitored. By contrast, the appearance of two transmission zeros in the frequency response indicates that different capacitive variations are introduced. Sensing based on measuring such notch frequencies is thus possible. It should be noted that any dielectric perturbation loaded to a single SISS can be sensed by monitoring the change in the resonance frequency given by (4.15). However, real-time differential measurements of two dielectric loads cannot be performed by loading a transmission line with a single SISS.

For differential sensing purposes, the differential input is the difference in the two capacitances (or relative permittivities of the loads), while the differential output is the difference between the two resonance frequencies (which may degenerate to the same one in the case of exactly equal capacitive perturbations). The differential sensitivity is defined as the variation in the difference between the resonance frequencies, f_d, due to the variation in the difference between the capacitances, C_{sd}, that is,

$$S = \frac{df_d}{dC_{sd}} = \frac{d(f_2 - f_1)}{d(C_{s2} - C_{s1})} = \lim_{\Delta(C_{s2}-C_{s1})\to 0} \frac{\Delta(f_2 - f_1)}{\Delta(C_{s2} - C_{s1})} \text{(Hz/F)}. \tag{6.9}$$

If one of the SISS is loaded with a sample material of reference (so that we may assume $\Delta C_{s1} = 0$), then (6.9) becomes

$$S = \lim_{\Delta C_{s2}\to 0} \frac{\Delta(f_2 - f_1)}{\Delta C_{s2}}. \tag{6.10}$$

From the equivalent circuit in Fig. 4.56b, we can gain insight into the theoretical resonance frequencies and sensitivity as a function of the input differential capacitance by numerical solutions (see Fig. 6.27). Unbalanced perturbations are introduced by a capacitive variation in one of the SISSs ($\Delta C_{s1} = 0$, $\Delta C_{s2} \neq 0$), and different coupling coefficients, $k_m = -|M|/L_s$,[25] are considered. As expected, for balanced perturbations ($\Delta C_{s1} = \Delta C_{s2} = 0$), the frequency response exhibits a single transmission zero at f_e given by (4.14). By contrast, when the capacitive perturbations are different from each other ($\Delta C_{s2} \neq 0$), two resonance frequencies predicted by (4.16) appear,[26] each of them being dependent on both capacitive perturbations. This means that the two resonance frequencies are not independent to one another (because of the existing resonator-to-resonator coupling), and a wide bandwidth is required since coupling enhances the separation between the resonance frequencies.

[25] The case with $k_m = 0$ cannot be implemented with a pair of SISSs sharing the same junction plane to the line, but this situation describes the case with isolated SISSs.

[26] When $k_m \neq 0$, the lower resonance frequency, f_1, is more sensitive to $+|\Delta C_{s2}|$ than the upper resonance frequency, f_2. Complementarily, f_2 is more sensitive to $-|\Delta C_{s2}|$. In these sensitive regions, the resonances tend to follow the resonance frequency of the perturbed uncoupled SISS (f_2 for $k_m = 0$).

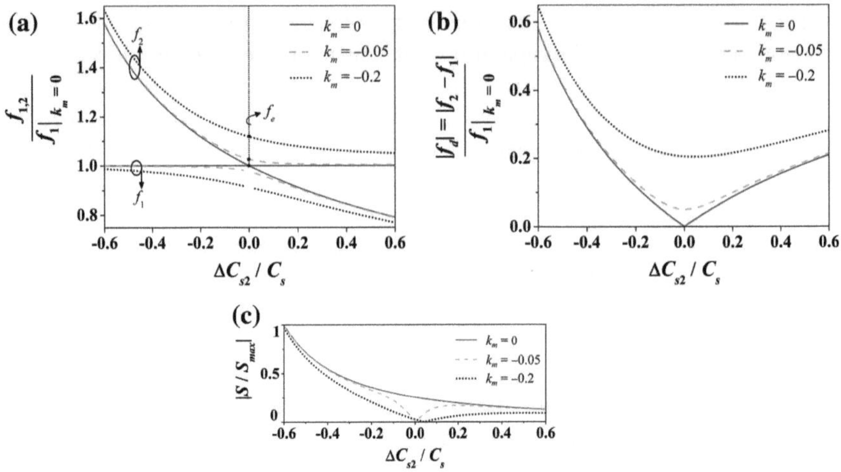

Fig. 6.27 Parallel-connected pair of identical SISSs loaded with the capacitive perturbations $\Delta C_{s1} = 0$ and $-0.6 \leq \Delta C_{s2}/C_s \leq 0.6$ for different k_m. **a** Resonance frequencies given by (4.14)–(4.16) normalized to the constant resonance frequency of the non-perturbed uncoupled SISS, i.e. f_1 with $k_m = 0$. **b** Magnitude of the normalized transfer function. **c** Magnitude of the sensitivity given by (6.10) normalized to its maximum value, S_{max}. Reprinted with permission from [28]

As the inter-resonator coupling strength ($|k_m|$) increases, the difference in the two resonance frequencies ($|f_2 - f_1|$) increases, degrading the sensitivity. Indeed, the maximum sensitivity corresponds to the hypothetical case of uncoupled SISSs (See footnote 26). It is important to highlight that a significant drop in the sensitivity occurs for small perturbations, even for small coupling coefficients.

The previous results suggest that the considered topology of parallel-connected SISSs is not much suitable to operate as a sensor due to the inevitably presence of inter-resonator coupling. On the other hand, the resonance frequency separation caused by coupling may be interesting to design a comparator. However, the smaller the difference in the perturbations, the narrower the lower notch (the bandwidth of the lower resonance is narrow by nature, as compared to that of the upper resonance). Therefore, the resolution for small differential inputs, which is of special interest for the design of comparators, is expected to be degraded by losses (the notch may be notably shallow giving rise to a poor resolution, or even not detected). Note also that since the transfer function is not linear, the sensitivity is not constant.

Although in the equivalent circuit model the perturbations are modeled by generalized capacitances, in a permittivity sensor the perturbations must be due to changes in the relative permittivity of dielectric loads (samples of the materials, substances or liquids under test). The optimum position of the dielectric loads to enhance sensitivity is where the electric energy is concentrated, namely, in the vicinity of the low impedance section of the SISSs (Sect. 4.4.1). An on-wafer in-house proof-of-concept prototype operating around 60 GHz is fabricated to experimentally validate the sensing principle using dielectric loading. The structure, which is shown in Fig. 6.28a, is

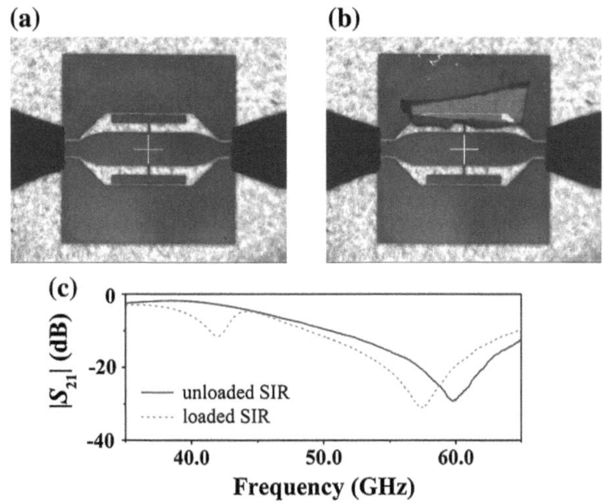

Fig. 6.28 Photograph of the proof-of-concept demonstrator based on CB-CPW **a** without and **b** with a piece of glass with $\epsilon_r = 4.6$ on top of one of the SISS. **c** Measured transmission coefficient. The dimensions are (see Fig. 4.55): $W = 350\ \mu m$, $G = 210\ \mu m$ (50-Ω line), $l_1 = 90\ \mu m$, $l_2 = 100\ \mu m$, $w_1 = 800\ \mu m$, and $w_2 = 20\ \mu m$. The substrate parameters are: $h = 200\ \mu m$, $\epsilon_r = 4.6$, and $\tan\delta = 0.0021$. Reprinted with permission from [28]

based on a mirrored pair of parallel SISSs implemented on conductor-backed coplanar waveguide (CB-CPW) technology (this technology is dictated by the available experimental set-up). Microwave CPW probes are used to measure the structure without external inclusions and with a piece of dielectric glass on top of one of the SISSs, as can be seen in Fig. 6.28b. As expected, in reference to Fig. 6.28c there is a single transmission zero when the structure is not loaded (the structure remains symmetric). By contrast, symmetry is disrupted by loading a dielectric sample in one of the resonators, giving rise to the appearance of two transmission notches. In other words, the lower transmission zero caused by symmetry perturbation validates experimentally the sensing principle.

Finally, it is worth mentioning that an alternative topology will be proposed in the near future in an expanded work of [28]. The topology is aimed at preventing inter-resonator coupling, and is intended to face with the drawbacks that parallel SISSs suffer. Specifically, the structure consists of a cascade connection of two identical SISSs, spaced by a transmission line section. The resonators are sufficiently separated so that their coupling is negligible. Thereby for asymmetric perturbations, the two resonance frequencies can be tuned independently because the coupling between the resonators is prevented. Preliminary results (not included in this book) show that a cascade connection of SISSs is a good solution to avoid coupling between them, achieving in turn significant advantages: the resonances are independent to each other (enhancing the sensitivity) and their bandwidth is not narrow by nature (improving the resolution), and narrower frequency bandwidth of operation is required since the resonance frequencies are not separated by coupling. Work is in progress to design and characterize practical differential permittivity sensors and comparators.

References

1. M. Schüßler, C. Mandel, M. Puentes, R. Jakoby, Metamaterial inspired microwave sensors. IEEE Microw. Mag. **13**(2), 57–68 (2012)
2. C. Mandel, B. Kubina, M. Schüßler, R. Jakoby, Metamaterial-inspired passive chipless radio-frequency identification and wireless sensing. Annals of telecommunications-annales des télécommunications **68**(7–8), 385–399 (2013)
3. C. Mandel, B. Kubina, M. Schüßler, R. Jakoby, Passive chipless wireless sensor for two-dimensional displacement measurement, in *European Microwave Confererrnce (EuMC)*, Manchester, UK, pp. 79–82, Oct. 2011
4. X. Liu, C. Xue, S. Yan, J. Xiong, W. Zhang, Integrated high sensitivity displacement sensor based on micro ring resonator, in *4th IEEE International Conference Nano/Micro Engineered and Molecular Systems (NEMS)*, Shenzhen, China, Jan. 2009
5. Z. Shaterian, A.K. Horestani, C. Fumeaux, Metamaterial-inspired displacement sensor with high dynamic range, in *4th International Conference Metamaterials, Photonic Crystals Plasmonics (META)*, Sharjah, UAE, Mar. 2013
6. A. Ebrahimi, W. Withayachumnankul, S. Al-Sarawi, D. Abbott, Metamaterial-inspired rotation sensor with wide dynamic range. IEEE Sens. J. **14**(8), 2609–2614 (2014)
7. M. Puentes, C. Weiss, M. Schüßler, R. Jakoby, Sensor array based on split ring resonators for analysis of organic tissues, in *IEEE MTT-S International Microwave Symposium Digest*, Baltimore, MD, USA, 2011
8. M. Puentes, M. Maasch, M. Schüßler, R. Jakoby, Frequency multiplexed 2-dimensional sensor array based on split-ring resonators for organic tissue analysis. IEEE Trans. Microw. Theory Tech. **60**(6), 1720–1727 (2012)
9. M. Puentes, *Planar Metamaterial Based Microwave Sensor Arrays for Biomedical Analysis and Treatment*, Springer Thesis (Springer, Switzerland, 2014)
10. A. Abduljabar, D. Rowe, A. Porch, D. Barrow, Novel microwave microfluidic sensor using a microstrip split-ring resonator. IEEE Trans. Microw. Theory Tech. **62**(3), 679–688 (2014)
11. A. Ebrahimi, W. Withayachumnankul, S. Al-Sarawi, D. Abbott, High-sensitivity metamaterial-inspired sensor for microfluidic dielectric characterization. IEEE Sens. J. **14**(5), 1345–1351 (2014)
12. W. Withayachumnankul, K. Jaruwongrungsee, A. Tuantranont, C. Fumeaux, D. Abbott, Metamaterial-based microfluidic sensor for dielectric characterization. Sens. Actuators A Phys. **189**, 233–237 (2013)
13. T. Chretiennot, D. Dubuc, K. Grenier, Optimized electromagnetic interaction microwave resonator/microfluidic channel for enhanced liquid bio-sensor, in *European Microwave Conference (EuMC)*, Nuremberg, Germany, Oct. 2013, pp. 464–467
14. A. Albishi, M.S. Boybay, O.M. Ramahi, Complementary split-ring resonator for crack detection in metallic surfaces. IEEE Microw. Wirel. Compon. Lett. **22**(6), 330–332 (2012)
15. M. Boybay, O. Ramahi, Non-destructive thickness measurement using quasi-static resonators. IEEE Microw. Wirel. Compon. Lett. **23**(4), 217–219 (2013)
16. C. Damm, M. Schussler, M. Puentes, H. Maune, M. Maasch, R. Jakoby, Artificial transmission lines for high sensitive microwave sensors, in *IEEE Sensors*, Christchurch, New Zealand, Oct. 2009, pp. 755–758
17. C. Damm, *Artificial Transmission Line Structures for Tunable Microwave Components and Microwave Sensors* (Shaker, Aachen, 2011)
18. J. Fraden, *Handbook of Modern Sensors: Physics, Designs, and Applications*, 3rd edn. (Springer, New York, 2004)
19. J.G. Webster, *The Measurement Instrumentation and Sensors Handbook* (CRC Press, Boca Raton, 1999)
20. E. Ekmekci, G. Turhan-Sayan, Multi-functional metamaterial sensor based on a broad-side coupled SRR topology with a multi-layer substrate. Appl. Phys. A **110**(1), 189–197 (2013)
21. N. Suwan, Investigation of RF direct detection architecture circuits for metamaterial sensor applications, Master Thesis, University of Waterlo, Oct. 2011

22. A.K. Horestani, J. Naqui, Z. Shaterian, D. Abbott, C. Fumeaux, F. Martín, Two-dimensional alignment and displacement sensor based on movable broadside-coupled split ring resonators. Sens. Actuator A Phys. **210**, 18–24 (2014)
23. C.R. Paul, *Introduction to Electromagnetic Compatibility*, 2nd edn. (Wiley, New York, 2006)
24. J. Naqui, M. Durán-Sindreu, F. Martín, Novel sensors based on the symmetry properties of split ring resonators (SRRs). Sensors—Special Issue Metamaterials for Sensing **11**(8), 7545–7553 (2011)
25. A.K. Horestani, C. Fumeaux, S. Al-Sarawi, D. Abbott, Displacement sensor based on diamond-shaped tapered split ring resonator. IEEE Sens. J. **13**(4), 1153–1160 (2013)
26. A.K. Horestani, D. Abbott, C. Fumeaux, Rotation sensor based on horn-shaped split ring resonator. IEEE Sens. J. **13**(8), 3014–3015 (2013)
27. A.K. Horestani, J. Naqui, D. Abbott, C. Fumeaux, F. Martín, Two-dimensional displacement and alignment sensor based on reflection coefficients of open microstrip lines loaded with split ring resonators. IET Electron. Lett. **50**(8), 620–622 (2014)
28. J. Naqui, C. Damm, A. Wiens, R. Jakoby, L. Su, F. Martín, Transmission lines loaded with pairs of magnetically coupled stepped impedance resonators (SIRs): Modeling and application to microwave sensors, in *IEEE MTT-S International Microwave Symposium Digest*, Seattle, WA, USA, 2014
29. J. Naqui, M. Durán-Sindreu, F. Martín, Alignment and position sensors based on split ring resonators. Sensors—Special Issue Ultra-Small Sensor Systems and Components **12**(9), 11 790–11 797 (2012)
30. J. Naqui, M. Durán-Sindreu, F. Martín, On the symmetry properties of coplanar waveguides loaded with symmetric resonators: Analysis and potential applications, in *IEEE MTT-S International Microwave Symposium Digest*, Montreal, Canada, 2012
31. D.M. Pozar, *Microwave Engineering*, 3rd edn. (Wiley, Hoboken, 2005)
32. J. Naqui, F. Martín, Mechanically reconfigurable microstrip lines loaded with stepped impedance resonators and potential applications. Int. J. Antennas Propag. **2014**(346838), 8 (2014)
33. V. Sipal, A. Narbudowicz, M. Ammann, Contactless measurement of angular velocity using circularly polarized antennas. IEEE Sens. J. **15**(6), 3459–3466 (2015)
34. J. Naqui, M. Durán-Sindreu, F. Martín, Transmission lines loaded with bisymmetric resonators and applications, in *IEEE MTT-S International Microwave Symposium Digest*, Seattle, WA, USA, 2013
35. J. Naqui, F. Martín, Transmission lines loaded with bisymmetric resonators and their application to angular displacement and velocity sensors. IEEE Trans. Microw. Theory Tech. **61**(12), 4700–4713 (2013)
36. J. Naqui, F. Martín, Angular displacement and velocity sensors based on electric-LC (ELC) loaded microstrip lines. IEEE Sens. J. **14**(4), 939–940 (2014)

Chapter 7
Application of Symmetry Properties to Spectral Signature Barcodes

The symmetry properties of transmission lines loaded with single (per-unit-cell) symmetric resonators described in Chap. 4 may be used not only as a sensing mechanism (Chap. 6), but also as an encoding mechanism. In this chapter, preliminary results applying such symmetry properties for the synthesis of chipless tags for radio-frequency identification (RFID) are shown. In particular, the considered resonator-loaded lines, whose behavior can be determined by symmetry-related properties, may be interesting for RF barcodes implemented with multiresonating structures that encodes in the frequency domain. A binary encoding is merely done by making use of a coupling-modulated resonance, that is, by switching from an all-pass response to a stopband response, depending on whether resonance frequency is prevented or not, respectively. We discuss some implementations, and their advantages and limitations are highlighted.

Section 7.1 presents a brief overview of RFID and multiresonator-based chipless tags. Section 7.2 is devoted to the encoding mechanism based on symmetry properties, which is the main contribution of this thesis to multiresonator-based chipless tags. Finally, Sects. 7.3 and 7.4 are addressed to the presented proof-of-concept (3-bit) barcodes in coplanar waveguide and microstrip technologies, respectively, that validate the proposed approach for encoding purposes.

7.1 Introduction

Radio-frequency identification (RFID) is a wireless data capturing technique that utilizes radio frequency (RF) waves for automatic identification of objects. RFID systems consist of three major components: a reader or interrogator, a tag or transponder, and middleware software. The reader sends interrogation signals to the tag that is required to be identified, and the software decodes the identification data. In recent

J. Naqui, *Symmetry Properties in Transmission Lines Loaded with Electrically Small Resonators*, Springer Theses,
DOI 10.1007/978-3-319-24566-9_7

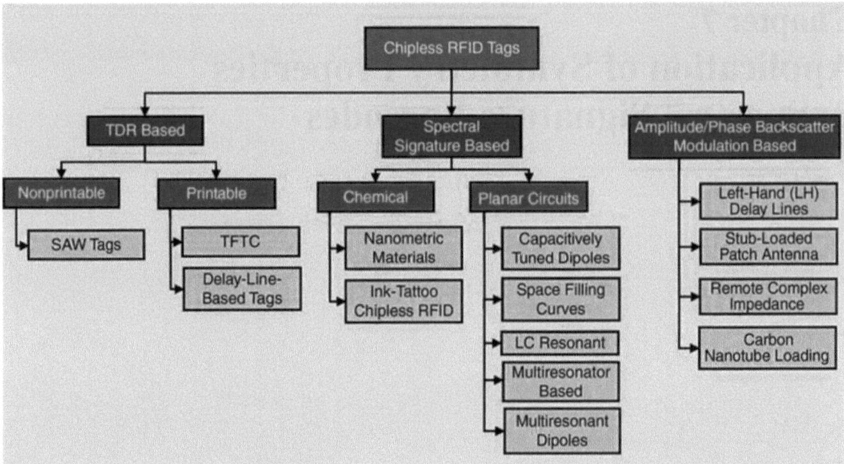

Fig. 7.1 Classification of chipless RFID tags. In this thesis, some proof-of-concept multiresonator-based spectral signature tags inspired by symmetry properties have been developed. Reprinted with permission from [1]

years the tendency is to replace conventional *optical barcodes* by RFID tags.[1] This trend is due to the fact that RFID offers several advantages over conventional barcodes, such as unique identification code for individual items, longer reading ranges, reading in the presence of obstacles, and automatic identification [1]. The only reason why RFID tags have not replaced the traditional optical barcode is the price of the tag, which is higher than the price of the optical barcode. *Passive tags* are those that do not have any on-board power supply. In addition, the tags are regarded as *chipless tags* if the passive tags do not contain any integrated circuit (IC) [1]. Chipless tags are aimed at reducing the tag price as much as possible, obviously at the expense of performance (such as a poorer data storage capability or a shorter reading range). Although chipless RFID tags are relatively recent, a number of approaches has been already developed, as is shown in Fig. 7.1. A comprehensive review of chipless RFID is out of the scope of this chapter, but it is available in the literature by Preradovic and Karmakar [1, 2].

In this thesis we are focused on multiresonator-based chipless RFID tags. This sort of chipless tags was designed by Preradovic et al. at Monash University [3], and it was patented in 2008 [4]. Shortly after, in 2009, a Ph.D. dissertation devoted to multiresonator-based chipless RFID systems was presented by Preradovic [5]. In fact, a large number of contributions on these chipless RFID systems has been published by the same authors. In these RF barcodes the information is stored in

[1]Nowadays optical devices are widely spread and can be found not only in identification systems, but also in sensors. Currently, many efforts are made in the microwave community to design identification/sensor systems with competitive performance in order to replace the optical devices with microwave devices.

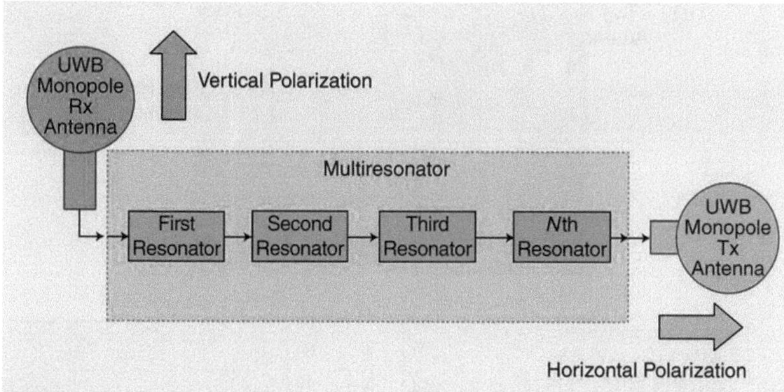

Fig. 7.2 Block diagram of multiresonator-based chipless tags for spectral signature encoding. Reprinted with permission from [1]

the so-called spectral signature of the tag (frequency-domain encoding). As is illustrated in Fig. 7.2, the chipless tag is composed of a multiresonating structure, and cross-polarized receiving and transmitting antennas. Therefore, interrogation and response signals are separated by changing the polarization, typically linear using ultra-wideband (UWB) monopoles. The spectral signature is obtained by loading a transmission line with an array of resonators, each one resonating at a different frequency. Thereby we can set a binary code, where each resonator corresponds to a bit value (logic '1' or '0') that is controlled by the presence or absence of the resonance frequency corresponding to that bit. The encoding is typically in amplitude, which is characterized by notches in the spectral signature, although phase encoding has also been reported. By interrogating the chipless tag with a multi-frequency signal, it is possible to detect the variations in the amplitude and phase of the received signal from the tag in order to decode the identification [5, 6]. In the literature, tags implemented in microstrip technology as well as in coplanar waveguide technology have been presented [1, 5, 7]. Figure 7.3 shows a multiresonator-based chipless tag reported by Preradovic et al. that may be regarded as a reference one [8].

As compared to other chipless tags, the advantages of the multiresonator-based approach are [1]: fully printable, low cost, and robust. Therefore, these tags can be implemented in flexible substrates by means of printed techniques for low-cost item tracking [9–11]. Conversely, this approach requires a large spectrum, a wideband reader, and an appropriate tag orientation (cross-polarized transmitting and receiving antennas are used) [1, 12, 13]. A key aspect in RFID is that the most suitable technology depends on the requirements of the application. The multiresonator-based chipless tags may be interesting for applications that require neither a high level of security nor to store too much information (e.g. for document identification or electoral processes with small voting members).

There are several considerations to bear in mind when designing multiresonator-based tags. The data storage capacity is related to the spectral band efficiency.

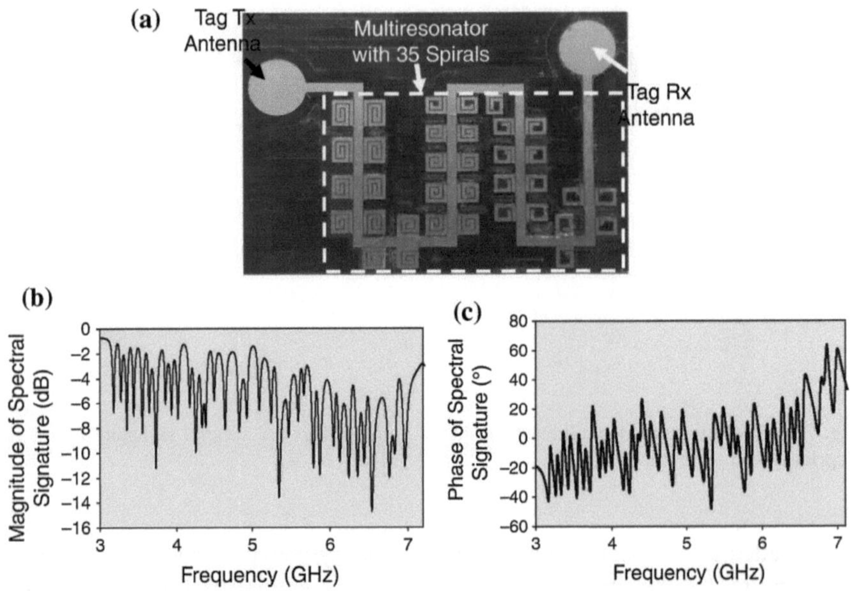

Fig. 7.3 35-bit chipless tag based on a microstrip line loaded with spiral resonators. **a** Photograph. **b** Magnitude and **c** phase of the spectral signature. Reprinted with permission from [1]

In order to accommodate the maximum number of resonance frequencies in a predefined frequency spectrum, notches should be as narrow as possible (typically weak line-to-resonator coupling is required). Even though the resulting notch magnitude (intimately related to the bandwidth) is not very deep, this is not a critical drawback in binary codes. Another requirement is that since the resonance frequencies of coupled resonators change with the coupling, inter-resonator coupling must be avoided to prevent resonance frequencies from shifting. Instead, the resonance frequencies should be independent to each other (i.e. each resonance should be tuned solely by its corresponding resonator). Evidently, there is a tradeoff because the resonators should be as close as possible with a view to reducing the tag size.

As mentioned in Chap. 6, there is an increasingly trend to develop microwave wireless sensors integrated with RFID [14–21]. An exhaustive overview on this topic is provided in [14].

7.2 Symmetry-Based Encoding

In the conventional approach [1–5], resonance is established (logic level '1' in the considered convention) by loading a transmission line with a resonator. On the contrary, the resonance is prevented (logic '0') by removing or short-circuiting the resonator (the latter shifts the resonance frequency upward and outside the frequency band of the spectral signature). In this book an alternative to this standard approach is proposed, whose encoding mechanism is based on exploiting the symmetry properties of transmission lines loaded with symmetric resonators explained in Chap. 4. More specifically, by virtue of a coupling-modulated resonance, it suffices to align properly a resonator to a transmission line in order to prevent resonance; the necessary and sufficient condition is that the symmetry planes of the line and resonator must be of different electromagnetic nature (one a magnetic wall and the other one an electric wall).

Therefore, the main advantage of the proposed approach using symmetric resonators is that it is not necessary to remove or short-circuit the resonators to set a logic '0'. The binary values can be simply modified to '1' by displacing or rotating the resonators, or by any other means of rupturing symmetry of the bit cells (e.g. by adding adhesives with metallic or dielectric loads). By considering that symmetry is broken by moving the resonators from the aligned position, interestingly, reconfigurable (programmable) barcodes are potentially possible as long as the resonators can be somehow shifted independently. The constraining challenge, however, concerns to how reconfigure mechanically the resonators. On the other hand, it can be readily observed that one limitation of the proposed encoding mechanism is that a transmission line can be loaded just with one (per-unit-cell) resonator. As stressed in Sect. 4.4, resonance may be inhibited provided that a transmission line is loaded with a single resonator at a specific transverse-oriented region. Accordingly, for a particular number of resonators, the tag cannot be as compact as using, for example, pairs of resonators (see the tag of Fig. 7.3).[2]

In this book, the focus is only on the multiresonating structure, which is characterized by its transmission coefficient (a full tag with antennas has not been designed yet). Next sections are devoted to the reported proof-of-concept multiresonating structures implemented with different transmission lines and resonators. Each multiresonating structure consists of a transmission line loaded with three resonators with different dimensions to provide three different resonance frequencies, so that each resonator stands for 1-bit information. The presented 3-bit frequency-coded barcodes are encoded in amplitude in the frequency range 2–3 GHz, and the inter-notch spectral separation is about 100 MHz. The encoding is exclusively proven by the alignment/misalignment of symmetric (or bisymmetric) resonators with a transmission line, as was explained in Sect. 4.3. Since losses play a fundamental role in the attenuation, losses must be taken into account to characterize the RF barcode performance.

[2]Note that since the notched response in the proposed symmetry-based approach depends on the level of asymmetry, it seems feasible the possibility of providing more than 1-bit information per resonator. In amplitude encoding, the higher the dynamic range in the notch depth, the higher the number of logic levels. However, a tradeoff arises in the spectral band efficiency because the bandwidth of the notch widens with the notch amplitude. It should be noted that this approach (not explored in this thesis) may be regarded as the digitalization of the sensors based on symmetry rupture using single resonators (Chap. 6).

7.3 Coplanar Waveguide Implementations

Coplanar waveguides (CPWs) are interesting transmission lines due to the fact that occupy one of the two metalization layers available on the substrate. This means that, besides the coplanar transmission line layer, there is a full layer for the etching of the resonators, regardless of the resonator type (strip or slot). By contrast, a general drawback of devices implemented in CPW technology is that they may suffer from mode conversion (from the CPW mode to the slot mode) if the structures are not symmetric. In these cases, vias and backside strips connecting the lateral ground planes are required to suppress the slot mode, which increase the fabrication complexity and cost as well.

7.3.1 SRR- and FSIR-Loaded Coplanar Waveguides

Barcodes of 3-bit capacity consisting of a CPW loaded with split-ring resonators (SRRs) or folded stepped-impedance resonators (FSIRs) are reported in [22] and [23], respectively. The former is shown in Fig. 7.4, whereas the latter in Fig. 7.5. The behavior of these topologies was explained in Sect. 4.3.1. The resonators are etched on the back side of the substrate, and are aligned or laterally misaligned with the line (specifically aligned with one of the CPW slots) to set the corresponding binary values. Adjacent resonators are laterally displaced in opposite directions in order to minimize inter-resonator coupling. It is apparent that barcodes based on these topologies are in general asymmetric structures (the only symmetric structure is that corresponding to the code with all the resonators lined up with the line). For the purpose of preventing the presence of the slot mode, via-holes in the substrate

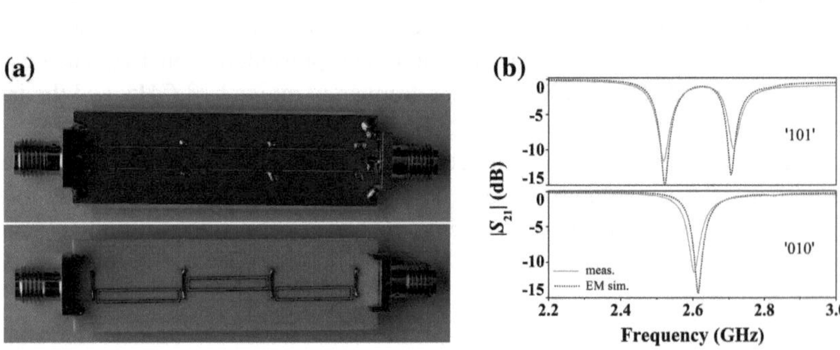

Fig. 7.4 3-bit barcode using **a** a CPW loaded with SRRs encoded as '101' and **b** transmission coefficient. The codification corresponding to the code '010' is also plotted. The dimensions are: for the CPW, $W = 3.3$ mm and $G = 0.2$ mm (50-Ω line); for the SRRs (see Fig. 4.6), $c = d = 0.2$ mm, $l_1(2.5\text{ GHz}) = 13.25$ mm, $l_1(2.6\text{ GHz}) = 13.75$ mm, $l_1(2.7\text{ GHz}) = 14.3$ mm, and $l_2 = 2.47$ mm; the longitudinal inter-resonator distance is $d_R = 1.15$ mm. The substrate is *Rogers RO4003C* with thickness $h = 0.8128$ mm, relative permittivity $\epsilon_r = 3.1$ (the nominal values are $\epsilon_r = 3.38$ and $\epsilon_r = 3.55$), and loss tangent $\tan\delta = 0.0021$. Reprinted with permission from [22]

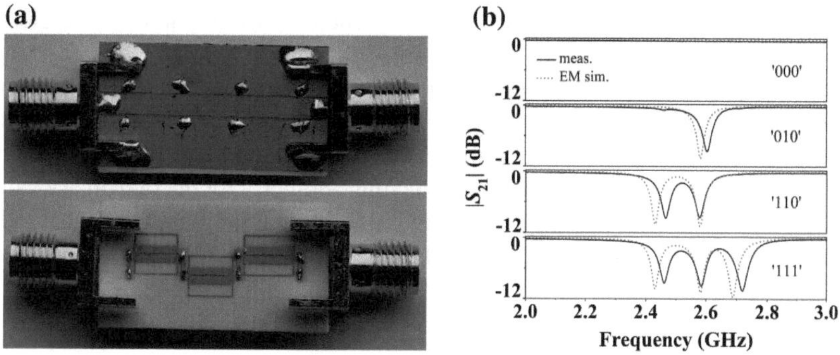

Fig. 7.5 **a** Fabricated CPW loaded with 3 FSIRs corresponding to the binary code '111' of the designed 3-bit RF barcode, and **b** transmission coefficient corresponding to the indicated codes. The substrate is *Rogers RO4003C* with $h = 0.8128$ mm, $\epsilon_r = 3.55$, and $\tan\delta = 0.0021$. CPW dimensions are: $W = 2.16$ mm, and $G = 0.15$ mm (50-Ω line); FSIR dimensions are (see Fig. 2.23): $l_1(2.5\text{ GHz}) = 5.55$ mm, $l_1(2.6\text{ GHz}) = 5.15$ mm, $l_1(2.7\text{ GHz}) = 4.9$ mm, $l_2 = 4.47$ mm, $W_1 = 0.825$ mm, $W_2 = 0.15$ mm, and $S = 0.15$ mm; $d_R = 1$ mm. Reprinted with permission from [23]

and strips on the bottom side of the substrate (on the same layer as the resonators) are used at the inter-resonator interfaces.

7.3.2 ELC-Loaded Coplanar Waveguides

The electric inductive-capacitive (ELC) resonator is an interesting particle to implement via-less barcodes in CPW technology. The reason lies on its bisymmetry, and to the application of a 90° rotation instead of a linear displacement as encoding

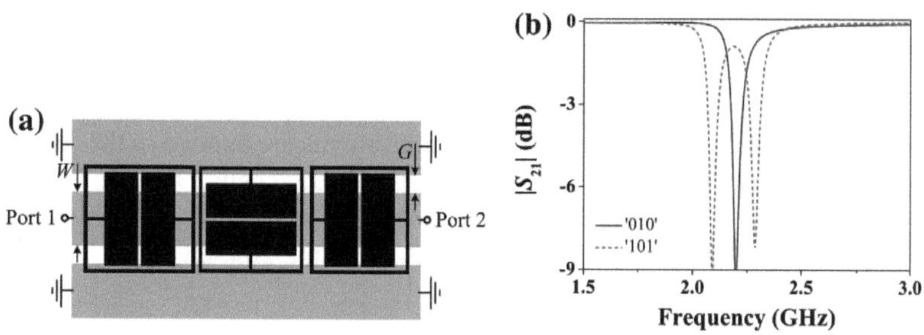

Fig. 7.6 **a** Layout of a 3-bit barcode encoded as '101' composed of a CPW loaded with ELC resonators, and **b** transmission coefficient for the indicated states obtained from lossy full-wave electromagnetic simulation. The substrate is *Rogers RO3010* with $h = 1.27$ mm, $\epsilon_r = 11.2$, and $\tan\delta = 0.0023$. Dimensions are: for the CPW, $W = 3$ mm and $G = 1$ mm (50-Ω line); for the ELC (see Fig. 2.25): $w_1 = 5.2$ mm, $w_2 = s = 0.2$ mm, $l_1 = 1.9$ mm, $l_2(2.1\text{ GHz}) = 6.32$ mm, $l_2(2.2\text{ GHz}) = 5.9$ mm, $l_2(2.3\text{ GHz}) = 5.62$ mm, and $l_3 = 6$ mm; $d_R = 0.2$ mm. Reprinted with permission from [24]

mechanism. As explained in Sect. 4.3.3, the structure is transparent if the electric wall of the resonator coincides with the symmetry plane of the line (state '0'). Then, by rotating the resonator 90°, symmetry is preserved (and thus the slot mode cannot emerge) while strong magnetic coupling between the line and the resonator arises, and the transmission coefficient exhibits a notch with maximum attenuation for this angle. It is apparent that the advantage of having symmetry for the two binary values is at the expense of a larger size as compared to SRR- or FSIR-based CPW barcodes. Figure 7.6 shows a 3-bit proof-of-concept barcode based on this topology presented in [24].

7.4 Microstrip Line Implementations

An important advantage of barcodes implemented through a microstrip line is that no parasitic modes are generated when the resonators are not lined up with the transmission line. In contrast to this, and as mentioned earlier, vias and strips are generally necessary in CPW barcodes to suppress the parasitic slot mode. Another advantage of microstrip implementation is that, as long as the ground plane is kept unaltered, the ground provides backside isolation so that objects made up of conductive materials can be labeled. To avoid ground plane etching, however, three metal layers are required if strips resonators are utilized (for ground, signal strip, and resonator). On the other hand, the disadvantage of the microstrip line is indeed this three-layer requirement owing to the presence of the ground plane.

7.4.1 FSIR-Loaded Microstrip Lines

Section 4.3.2 dealt with microtrip lines loaded with SIRs [25]. References [24, 25] present a 3-bit barcode based on symmetry properties implemented by loading folded SIRs (FSIRs) on a third metallic layer situated above the signal strip (see the cross section in Fig. 4.35). Figure 7.7 shows the top view for a specific code and the spectral signature for two different codes. Lateral displacements are applied to the resonators with a view to breaking the symmetry of the individual bit cells. Because the ground plane is kept unaltered, this implementation provides backside isolation by means of it, but the resonators must be etched on an additional metallic layer.

7.4.2 MLC-Loaded Microstrip Lines

Reference [26] is devoted to the analysis of microstrip lines loaded with magnetic inductive-capacitive (MLC) resonators (Sect. 4.3.4). As a possible application, this article presents a 3-bit barcode with the electric wall of the MLC aligned or mis-

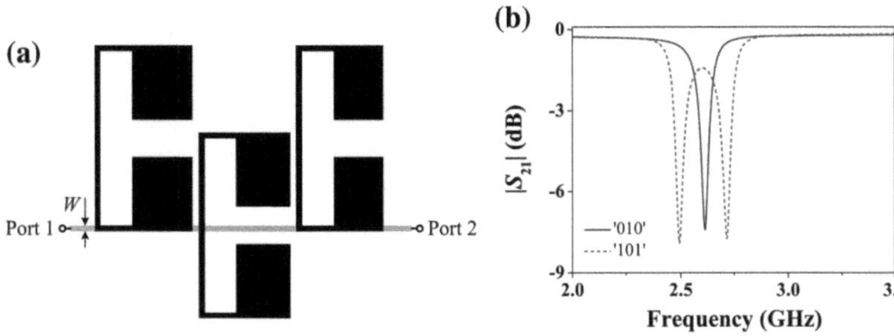

Fig. 7.7 **a** Microstrip line loaded with FSIRs for the code '101', and **b** lossy electromagnetic simulation of the transmission coefficient corresponding to the indicated codes. The top and bottom substrates (see Fig. 4.35) are *Rogers RO3010* with $h = 0.254$ mm, $\epsilon_r = 11.2$, and $\tan\delta = 0.0023$. The dimensions are: line width $W = 0.2$ mm (50-Ω line); for the FSIRs (see Fig. 2.23): $w_1(2.5\ GHz) = 2$ mm, $w_1(2.6\ GHz) = 1.8$ mm, $w_1(2.7\ GHz) = 1.64$ mm, $w_2 = 0.2$ mm, $s = 1.2$ mm, $l_1(2.5\,\text{GHz}) = 3.26$ mm, $l_1(2.6\,\text{GHz}) = 3.06$ mm, $l_1(2.7\,\text{GHz}) = 2.9$ mm, and $l_2 = 6$ mm; $d_R = 0.2$ mm. Reprinted with permission from [25]

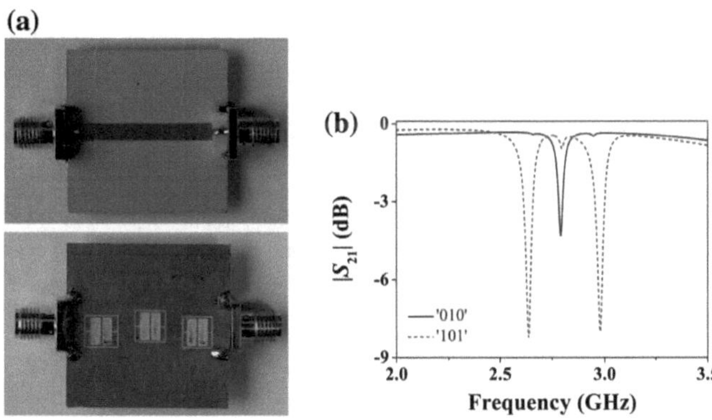

Fig. 7.8 **a** Photograph of a 3-bit RF barcode (encoded as '101'), and **b** measured frequency response. For completeness, the measurement for the code '010' is also shown. The substrate is *Rogers RO3010* with $h = 1.27$ mm, $\epsilon_r = 11.2$, and $\tan\delta = 0.0023$. The line width is $W = 3.13$ mm (by excluding the slot resonators, the characteristic impedance is 28 Ω). The MLC dimensions are (see Fig. 2.25): $w_1 = 5.2$ mm, $w_2 = s = 0.2$ mm, $l_1 = 1.9$ mm, $l_2(2.6\,\text{GHz}) = 6.32$ mm, $l_2(2.8\,\text{GHz}) = 6$ mm, $l_2(3\,\text{GHz}) = 5.72$ mm, and $l_3 = 6$ mm; $d_R = 3$ mm. The lateral displacement to set the logic state '1' is 1 mm. Reprinted with permission from [26]

aligned with the line, depending on the bit value to be implemented (see Fig. 7.8). This topology is the only one presented in this thesis that the resonators are etched in the ground plane, and it was found that this complicates the synthesis of the structure. One of the problems is that, as the MLC influences the characteristic impedance of the microstrip line to a considerable extent (Sect. 4.2.3), the line width is designed to minimize the insertion loss with the presence of the MLCs symmetrically printed. Another design difficulty is that contiguous resonators need to be sufficiently spaced apart to avoid mutual interaction, and the distance between adjacent resonators is set to that for which inter-resonator coupling can be neglected. Furthermore, as compared to SIR/FSIR-loaded microstrip lines, this approach does not provide backside isolation since the ground plane is etched, although it utilizes two metal layers.

References

1. S. Preradovic, N. Karmakar, Chipless RFID: bar code of the future. IEEE Microw. Mag. **11**(7), 87–97 (2010)
2. S. Preradovic, N.C. Karmakar, *Multiresonator-based Chipless RFID: Barcode of the Future*. (Springer Science & Business Media, New York, 2012)
3. S. Preradovic, I. Balbin, N. Karmakar, G. Swiegers, A novel chipless RFID system based on planar multiresonators for barcode replacement, in *IEEE International Conference on RFID*, (Las Vegas, NV, USA, 2008) pp. 289–296
4. S. Preradovic, I. Balbin, S. Roy, N. Karmakar, G. Sweigers, Radio frequency transponder, in *Australian Provisional Patent Application (Ref: P30228AUP1)*, pp. 79–82 (2008)
5. S. Preradovic, Chipless RFID system for barcode replacement. Ph.D. dissertation, Monash University, 2009
6. R. Koswatta, N. Karmakar, A novel reader architecture based on UWB chirp signal interrogation for multiresonator-based chipless RFID tag reading. IEEE Trans. Microw. Theory Tech. **60**(9), 2925–2933 (2012)
7. Y. Weng, S. Cheung, T. Yuk, L. Liu, Design of chipless UWB RFID system using a CPW multi-resonator. IEEE Antennas Propag. Mag. **55**(1), 13–31 (2013)
8. S. Preradovic, I. Balbin, N. Karmakar, G. Swiegers, Multiresonator-based chipless RFID system for low-cost item tracking. IEEE Trans. Microw. Theory Tech. **57**(5), 1411–1419 (2009)
9. L. Yang, A. Rida, R. Vyas, M. Tentzeris, RFID tag and RF structures on a paper substrate using inkjet-printing technology. IEEE Trans. Microw. Theory Tech. **55**(12), 2894–2901 (2007)
10. L. Zheng, S. Rodríguez, L. Zhang, B. Shao, L.-R. Zheng, Design and implementation of a fully reconfigurable chipless RFID tag using inkjet printing technology, in *IEEE International Symposium on Circuits System (ISCAS)*, (Seattle, WA, USA, 2008), pp. 1524–1527
11. S. Preradovic, N. Karmakar, Design of chipless RFID tag for operation on flexible laminates. IEEE Antennas Wireless Propag. Lett. **9**, 207–210 (2010)
12. B. Nikfal, C. Caloz, Hybrid time-frequency RFID system, in *IEEE MTT-S International Microwave Symposium Digest*, (Montreal, Canada, 2012)
13. N.C. Karmakar, R. Koswatta, P. Kalansuriya, E. Rubayet *et al.*, *Chipless RFID Reader Architecture*. (Artech House, Boston, 2013)
14. C. Mandel, B. Kubina, M. Schüßler, R. Jakoby, Metamaterial-inspired passive chipless radio-frequency identification and wireless sensing. *Annals of Telecommunications-annales des télécommunications* **68**(7-8), 385–399, (2013)
15. D. Girbau, A. Lázaro, R. Villarino, Passive chipless wireless sensor for two-dimensional displacement measurement, in *European Microwave Conference (EuMC)*, (Manchester, UK, 2011), pp. 79–82
16. D. Girbau, A. Lázaro, R. Villarino, Passive wireless permittivity sensor based on frequency-coded chipless RFID tags, in *IEEE MTT-S International Microwave Symposium on Digest*, (Montreal, Canada, 2012)
17. J. Lorenzo, D. Girbau, A. Lázaro, R. Villarino, Temperature sensor based on frequency-coded chipless RFID tags. Microw. Opt. Technol. Lett. **56**(10), 2411–2415 (2014)
18. D. Girbau, J. Lorenzo, A. Lázaro, C. Ferrater, R. Villarino, Frequency-coded chipless RFID tag based on dual-band resonators. IEEE Antennas Wireless Propag. Lett. **11**, 126–128 (2012)
19. T. Thai, J. Mehdi, F. Chebila, H. Aubert, P. Pons, G. DeJean, M. Tentzeris, R. Plana, Design and development of a novel passive wireless ultrasensitive RF temperature transducer for remote sensing. IEEE Sensors J. **12**(9), 2756–2766 (2012)

20. E.M. Amin, N. Karmakar, Development of a chipless RFID temperature sensor using cascaded spiral resonators, in *IEEE Sensors*, (Limerick, Ireland, 2011), pp. 554–557
21. S. Preradovic, N. Kamakar, E.M. Amin, Chipless RFID tag with integrated resistive and capacitive sensors, in *Asia-Pacific Microwave Conference Proceeding (APMC)*, pp. 1354–1357 (2011)
22. J. Naqui, M. Durán-Sindreu, F. Martín, Selective mode suppression in coplanar waveguides using metamaterial resonators. Appl. Phys. A **109**(4), 1053–1058 (2012)
23. J. Naqui, M. Durán-Sindreu, F. Martín, On the symmetry properties of coplanar waveguides loaded with symmetric resonators: Analysis and potential applications, in *IEEE MTT-S International Microwave Symposium Digest*, (Montreal, Canada, 2012)
24. J. Naqui, M. Durán-Sindreu, F. Martín, Strategies for the implementation of sensors and RF barcodes based on transmission lines loaded with symmetric resonators, in *21st International Conference Applied Electromagnetic Communication (ICECom)*, (Dubrovnik, Croatia, 2013)
25. J. Naqui, F. Martín, Mechanically reconfigurable microstrip lines loaded with stepped impedance resonators and potential applications. Int. J. Antennas Propag. **2014**(346838), (2014)
26. J. Naqui, M. Durán-Sindreu, F. Martín, Differential and single-ended microstrip lines loaded with slotted magnetic-LC (MLC) resonators. Int. J. Antennas Propag. **2013**(640514), 8 (2013)

Chapter 8
Conclusions and Future Work

8.1 General Conclusions

The main contribution reported in this thesis is the analysis, circuit modeling, and synthesis of transmission lines loaded with electrically small resonators (ESRs) in terms of symmetry considerations. It has been shown that the stopband characteristics (frequency, bandwidth, and/or attenuation) can be controlled as a function of symmetry-related electromagnetic properties. The theory derived from symmetry arguments is exploited to filtering, sensing, and encoding purposes. In particular, on the basis of symmetry-related properties, common-mode suppressed differential lines, microwave sensors, and radio frequency (RF) barcodes have been implemented.

The resonator-loaded lines have been implemented in planar technology using either microstrip or coplanar waveguide transmission lines, and some different topologies of planar resonators have been considered (most of them may be regarded as metamaterial resonators). The synthesis of these structures involves the choice of the most appropriate pair of line and resonator as well as the optimization of the geometrical topology depending on the specifications dictated by the application. A significant number of proof-of-concept prototypes support the validity and potentiality of the proposed symmetry-based approaches.

An important part of this thesis concerns the modeling by means of lumped-element equivalent circuit models. As is well known, there are numerous motivations to model electromagnetic structures with reliable equivalent circuit models. For instance, they provide an intuitive abstraction of the behavior of physical structures, they facilitate design and optimization stages, and their frequency- or time-domain response can be obtained practically instantaneously using circuit simulators. Throughout the development of this thesis (Chaps. 3–7), lumped-element equivalent circuit models of the considered electromagnetic structures have been proposed. The extraction of circuit parameters has been performed using available methodologies in the literature or proposing new procedures. The circuit model validity is restricted up to beyond the fundamental resonance frequency of the resonators (i.e. the region

J. Naqui, *Symmetry Properties in Transmission Lines Loaded with Electrically Small Resonators*, Springer Theses,
DOI 10.1007/978-3-319-24566-9_8

of interest). However, the accuracy of the models depends strongly on the electrical size of the unit cells (the dimensions of the cells are needed to be small relative to the wavelength). Therefore, verification is always required and, to this end, modern full-wave electromagnetic simulators are very useful. Even though these simulators are time consuming and computationally intensive, these simulators are highly accurate and excellent for verification purposes. Thus, throughout this thesis, the circuit simulations, the electromagnetic simulations, and the experimental results are in good agreement. This somehow validates the proposed equivalent circuit models in the sense that good agreement is always a necessary condition, but it does not guarantee that the equivalent circuit models are certainly valid.

8.2 Line-to-Resonator Magnetoelectric Coupling

Traditionally, mixed coupling is neglected in situations where the resonators may be actually both electrically and magnetically excited, so that only the dominant electric or magnetic coupling is considered. However, it is shown that depending on the arbitrary relative orientation between the transmission line and the resonators, mixed coupling must be considered to accurately model the structures. A relevant implication is that the available circuit parameter extraction procedures to date (which consider the dominant coupling only), that have been used for several years, may fail in the presence of mixed coupling. The reason is due to the fact that these methods are based on the mapping of the scattering parameters obtained by electromagnetic simulation, and these parameters may be influenced by mixed coupling. In order to deal with this issue, mixed magnetoelectric coupling in the line-to-resonator interaction, that directly depends on the relative orientation of the resonant elements, has been modeled for the first time (Chaps. 3 and 4).

It is of special relevance the particular case where the resonators exhibit a single symmetry plane. Transmission lines loaded with single resonators possessing a single symmetry plane cannot be symmetric along the midplanes in the longitudinal and transverse directions simultaneously. Thus, these structures can be symmetric in regard to the ports only when the symmetry plane of the resonators is oriented orthogonally to the line. For all the other resonator orientations, in essence, the phases of the reflection coefficients are different from each other. This is explained by the magnetic coupling or, more precisely, by the magnetic coupling in the presence of the electric coupling (and therefore mixed coupling). The orientation that favors mixed coupling is the one where the resonators are symmetrically aligned with the line (i.e. the structures are symmetric with respect to the longitudinally-oriented midplane). Indeed, these symmetric topologies are of interest in the design of devices based on symmetry properties, and for this reason it has been important to gain insight into their behavior.

8.3 Inter-Unit-Cell Inter-Resonator Coupling

It is well known that transmission lines periodically loaded with resonators (uncoupled to one another) exhibit a region of evanescent modes. With the assumption that the resonators are closely-spaced apart, a resonator-loaded transmission line with inter-resonator coupling has been modeled by a four-port circuit (Chap. 3). Two ports correspond to the conventional transmission line ports, while the other two additional ports are like artificial ports accounting for the inter-resonator coupling.

Considering an infinite periodic structure, we have applied Bloch-mode theory to the four-port network using the transmission (*ABCD*) matrix method. It has been shown that the considered structures may support bimodal propagation; evanescent, forward, backward, and complex modes are found to exist. The appearance of complex modes enhances the stopband bandwidth (at the lower edge of the region of evanescent modes), and they are interpreted as the destructive interference between forward and backward waves. The forward waves come from transmission line-type propagation, whereas the backward ones are propagated through the coupled resonators. Specifically, backward electro- or magneto-inductive waves are supported by chains of electrically or magnetically coupled resonators, respectively. Thus, inter-resonator coupling in the cell-to-cell interaction is absolutely mandatory for complex modes to emerge.

The analytical modal solutions inferred from the equivalent circuit models have been corroborated by obtaining the modal solutions through a full-wave eigenmode solver of the corresponding electromagnetic structures. Experimental evidence of rejection bandwidth enhancement due to inter-resonator coupling is also provided.

8.4 Synthesis of Electrically Small Resonators

In contrast to the analysis of electromagnetic structures where circuit elements of equivalent circuit models are extracted, it has also been important to synthesize physical topologies to optimize as much as possible the required specifications dictated by the corresponding application. In this regard, the limits on the synthesis of ideal lumped shunt-connected series resonators by means of microstrip stepped-impedance shunt-stubs have been investigated comprehensively (Chap. 3). We have provided the range of physically implementable values of inductance and capacitance, which is clearly finite in practice.

8.5 Symmetry Properties

For the purpose of analyzing the symmetry properties of resonator-loaded transmission lines, the fundamental resonance frequency of electrically small resonators (Chap. 2) and the fundamental propagation modes of transmission lines (Chap. 4) are studied subjected to symmetry considerations (Chap. 4). Because of symmetry and the electromagnetic excitation, we have characterized these modes in terms of electric or magnetic wall boundary conditions at their symmetry planes.

In transmission lines loaded with single symmetric resonators, regardless of the line-to-resonator coupling mechanism (electric and/or magnetic), a necessary condition to prevent the excitation of the resonance frequency has been established. If the symmetry planes of the line and resonator are aligned and are of distinct electromagnetic nature (one being an electric wall and the other one a magnetic wall), then there is no net mutual electromagnetic coupling capable of exciting the resonance. As a consequence, the structure may be viewed as a transmission line with an all-pass response. By contrast, resonance is established if the symmetry planes of the two elements are aligned but are of the same electromagnetic nature (either electric or magnetic walls), if the symmetry planes are misaligned, or if deviation from the ideal reference symmetry (with different walls aligned) is produced by any other means. Under these circumstances the structure exhibits a stopband (or notched) response. The interesting feature is that the stopband characteristics (frequency, bandwidth, and/or attenuation level) in the vicinity of the resonance frequency depend on the asymmetry respect to the reference symmetric case, since this asymmetry determines the net mutual coupling between the line and the resonator. Accordingly, transmission lines loaded with symmetric resonators may be said to be based on a *coupling-modulated resonance*.

In virtue of a coupling-modulated resonance, in symmetric coplanar waveguides and pairs of microstrip lines (three-conductor transmission lines), we have selectively suppressed (stopband) either the even- or the odd-mode while preserving the other mode almost unaltered (all-pass) by properly aligning symmetric resonators with the line (Chap. 4). This functionality is referred to as *selective mode suppression*, where the line-to-resonator coupling is null for one mode (for perfectly overlapped symmetry planes of different nature) whereas it is tight for the other mode (for perfectly overlapped symmetry planes of equal nature). The well-known *even/odd-mode analysis* has been applied in order to simplify the equivalent circuits, and the resulting even- and odd-mode networks have supported the theory relative to the selective mode suppression.

On the other hand, regarding transmission lines loaded with more than one resonator (placed transversely to the line), to have symmetry, identical mirrored resonators are necessary rather than symmetric resonators. The prevention of resonance in these structures is not possible because one cannot align more than one resonator with the line simultaneously. Nevertheless, transmission lines loaded with pairs of resonators have also been characterized in terms of symmetry, where controllability in the resonance frequencies of the structure is achieved. Particularly, in general there is either a single resonance frequency or two resonance frequencies, depending on whether the structure is symmetric or not through the midplane extending along the longitudinal direction, respectively. For such a reason, the stopband functionality in transmission lines loaded with multiple resonators (placed in the transverse direction to the line) is based on *resonance frequency splitting/shifting*.

In summary, stopband characteristics in resonator-loaded transmission lines have been controlled by symmetry properties, where the electromagnetic behavior is based on either a coupling-modulated resonance or resonance frequency splitting/shifting. From a coupling-modulated resonance, the signal filtering functionality termed selective mode suppression is derived.

8.6 Common-Mode Suppressed Differential Lines

As a practical application of the selective mode suppression functionality, *differential microstrip lines with common-mode rejection* have been proposed (Chap. 5).

A simple and systematic design procedure has been reported. With a view to enhancing the common-mode rejection bandwidth, two strategies have been investigated: (*i*) the design of unit cells exhibiting wideband common-mode rejection, and (*ii*) to couple the resonators as much as possible. It has been assumed periodic structures to allow for their characterization using Bloch-mode theory applied to the four-port equivalent circuit for the even mode (common mode). By that means the limits on the maximum rejection bandwidth obtained by periodicity may be easily inferred.

Some prototypes have been characterized and compared to devices based on other approaches; the reported solution is technologically simple, low cost, compact, provides wide common-mode stopbands with high rejection, and the differential-mode signal integrity is kept almost unaltered in a broad band. If necessary, further bandwidth enhancement can be obtained tuning the resonators at different frequencies.

8.7 Microwave Sensors

Novel microwave sensor concepts inspired by symmetry disruption have been introduced (Chap. 6). The operation principle of the proposed sensors are based on either a coupling-modulated resonance or resonance frequency splitting/shifting.

Regarding sensors based on a coupling-modulated resonance, proof-of-concept demonstrators of *spatial sensors* for alignment, displacement, position, and velocity measurements have been reported. These sensors consist of two parts that must be etched on different substrates (or objects), namely, a transmission line and a symmetric resonator. The relative position between them is estimated from the transmission coefficient (notch amplitude and/or frequency). The reason is that the transmission characteristics (stopband or all-pass) are strongly modulated (or controlled) by the relative orientation of the resonator with regard to the line, which determines the effective line-to-resonator coupling nature (electric and/or magnetic) and strength. Since these sensors can be *contactless*, they do not suffer from aging effects due to mechanical friction.

With respect to sensors whose sensing principle is based on resonance frequency splitting/shifting, an approach for the design of *differential sensors and comparators for permittivity measurements* has been proposed. These devices are composed of a transmission line loaded with pairs of identical and mirrored resonators. Under symmetry, the structure produces a single transmission zero. By contrast, when symmetry is broken, in general two transmission zeros appear which can be potentially used for sensing purposes. Therefore, a differential measurement may be performed by loading each resonator with a dielectric sample. The difference in the permittivity of the two samples is expected to be found by measuring the resonance frequencies of the structure. Preliminary experimental results have supported the sensing principle, although (intra-unit-cell) inter-resonator coupling, which is seen to exist in practice in the topology studied, should be canceled in order to enhance sensitivity by proposing alternative topologies.

In conclusion, the proposed spatial and permittivity sensors are simple (from the design viewpoint), compact in size, low cost (provided that the feeding and read-out systems are implemented by simple circuits and components), highly integrable, scalable, and robust against random errors produced by variable environmental conditions since differential measurements are performed based on geometrical symmetry which is not altered by those factors.

8.8 Spectral Signature Barcodes

Proof-of-concepts of spectral signature barcodes for radio-frequency identification (RFID) chipless tags inspired by symmetry properties have also been presented (Chap. 7). Spectral signature-based barcodes have been known for a long time. Particularly, in frequency-coded barcodes based on *multiresonating structures*, a host transmission line is loaded with multiple resonators tuned at different frequencies, each one contributing with one bit to the binary code; the presence or absence of a certain resonance gives the logic binary value (logic '1' or '0'). In the conventional approach, the resonators are removed or short-circuited to avoid their corresponding resonance frequencies. By contrast, in the reported strategy, the logic states are simply set by aligning or misaligning the resonators with the line (as suggested and argued, other symmetry disruption mechanisms can be alternatively used). We have discussed some implementations, highlighting their advantages and limitations.

8.9 Future Work

A considerable amount of further work may be carried out on the research lines emerged from this thesis. Among them, it should be highlighted:

- Investigate the potentiality of designing a balanced filter with common-mode suppression based on the resonant-type approach of metamaterial transmission lines (i.e. employing a combination of split-ring resonators and planar inductors/capacitors).
- Design of an angular velocity sensor suitable for space applications, and specifically for inertial wheels of satellites, where devices operate in harsh environments (Project A00016166 Innovation Triangle Initiative—ITI—). The aim is to replace the conventional optical encoders with microwave sensors.
- Integrate a simple measurement circuitry to the proposed sensing resonator-loaded transmission lines (Project A00016166 Innovation Triangle Initiative—ITI—).
- Development of a full chipless RFID tag (with transmitting and receiving antennas) for electronic voting using multiresonating structures inspired by symmetry-related properties with a competitive storage capacity using low-cost printing technologies and substrates (Project RTC-2014-2550-7 Retos de Colaboración 2014).
- Design and characterization of differential permittivity sensors and comparators based on pairs of stepped-impedance shunt-stubs or other planar resonators.
- Development of microwave wireless sensors integrated with RFID.

Appendix A
Dispersion Relation of Two-Port Networks

Consider an infinite structure composed of a cascade of identical two-port networks. Using an order-2 transmission ($ABCD$) matrix, we can relate the voltages and currents on either side of the unit cells, as illustrated in Fig. A.1, by

$$\begin{pmatrix} V_L \\ I_L \end{pmatrix} = \begin{pmatrix} A & B \\ C & D \end{pmatrix} \begin{pmatrix} V_R \\ I_R \end{pmatrix}. \tag{A.1}$$

For an infinite periodic structure, the variables on the right can differ from the variables on the left only by the propagation factor, $e^{\gamma l}$, where $\gamma = \alpha + j\beta$ is the complex propagation constant (α and β being the attenuation and phase constants, respectively) and l is the unit-cell length (βl being the electrical length). Accordingly, the above matrix equation can be written as [1, 2]

$$\begin{pmatrix} V_L \\ I_L \end{pmatrix} = \begin{pmatrix} A & B \\ C & D \end{pmatrix} \begin{pmatrix} V_R \\ I_R \end{pmatrix} = \begin{pmatrix} V_R \\ I_R \end{pmatrix} e^{\gamma l}. \tag{A.2}$$

The solution for the propagation constant is given by

$$\left(\begin{pmatrix} A & B \\ C & D \end{pmatrix} - \boldsymbol{I} e^{\gamma l} \right) \begin{pmatrix} V_R \\ I_R \end{pmatrix} = \begin{pmatrix} A - e^{\gamma l} & B \\ C & D - e^{\gamma l} \end{pmatrix} \begin{pmatrix} V_R \\ I_R \end{pmatrix} = 0, \tag{A.3}$$

where $\boldsymbol{I}$ is the identity matrix. For a non-trivial solution

$$\det \begin{pmatrix} A - e^{\gamma l} & B \\ C & D - e^{\gamma l} \end{pmatrix} = 0, \tag{A.4}$$

which yields

$$\cosh(\gamma l) = \frac{A + D}{2}. \tag{A.5}$$

J. Naqui, *Symmetry Properties in Transmission Lines Loaded with Electrically Small Resonators*, Springer Theses,
DOI 10.1007/978-3-319-24566-9

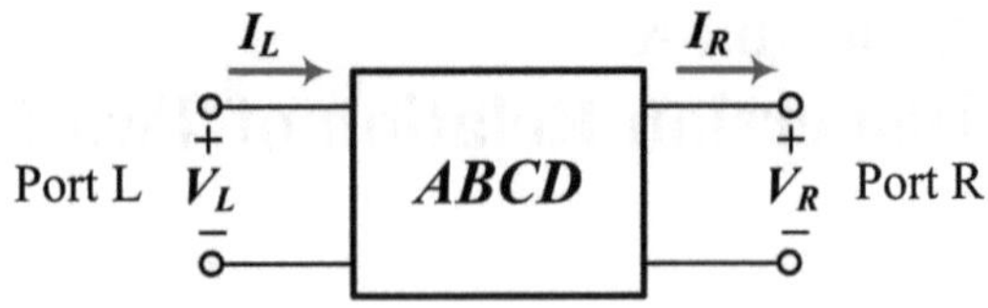

Fig. A.1 Two-port network characterization by the *ABCD* matrix

For lossless networks and propagating modes ($\alpha = 0$), we can replace $\cosh(\gamma l)$ with $\cos(\beta l)$, and hence the dispersion relation can be rewritten as

$$\cos(\beta l) = \frac{A + D}{2}. \tag{A.6}$$

In addition, for a symmetric network with regard to the ports, $A = D$, so that the solution reduces to

$$\cos(\beta l) = A. \tag{A.7}$$

References

1. D.M. Pozar, *Microwave Engineering*, 3rd edn. (Wiley, New York, 2005)
2. C. Caloz, T. Itoh, *Electromagnetic Metamaterials: Transmission Line Theory and Microwave Applications* (Wiley, New York, 2005)

Appendix B
Equivalent Circuit Models of Electric and Magnetic Couplings

B.1 Electric Coupling

Figure B.1a shows a lumped-element equivalent circuit model for a pair of capacitors, C_1 and C_2, which are coupled through a mutual capacitance C_m. The electric coupling coefficient is defined as [1, 2]

$$k_e = \frac{C_m}{\sqrt{C_1 C_2}}. \tag{B.1}$$

In the frequency domain, for sinusoidal time-dependence signals, the currents and voltages at the ports are related by

$$I_1 = j\omega C_1 V_1 - j\omega C_m V_2, \tag{B.2a}$$

$$I_2 = j\omega C_2 V_2 - j\omega C_m V_1. \tag{B.2b}$$

By connecting the lower terminals of the circuit model to form a three-terminal network, an equivalent π-circuit composed of only capacitances can be obtained, as depicted in Fig. B.1b [1, 2]. Accordingly, electric coupling is usually modeled by a capacitor connecting the coupled elements.

B.2 Magnetic Coupling

Figure B.2a illustrates a lumped-element equivalent circuit model of two inductors, L_1 and L_2, coupled by the mutual inductance L_m. The magnetic coupling coefficient is expressed as [1, 2]

$$k_m = \frac{L_m}{\sqrt{L_1 L_2}}. \tag{B.3}$$

J. Naqui, *Symmetry Properties in Transmission Lines Loaded with Electrically Small Resonators*, Springer Theses,
DOI 10.1007/978-3-319-24566-9

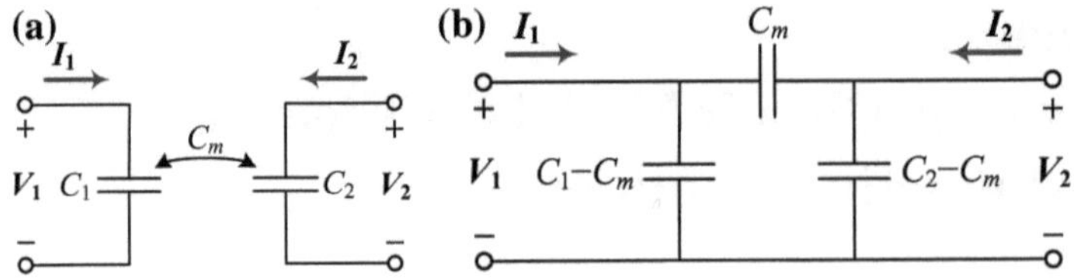

Fig. B.1 **a** Equivalent circuit model of capacitive coupling between two capacitors through a mutual capacitance and **b** transformed equivalent π-circuit

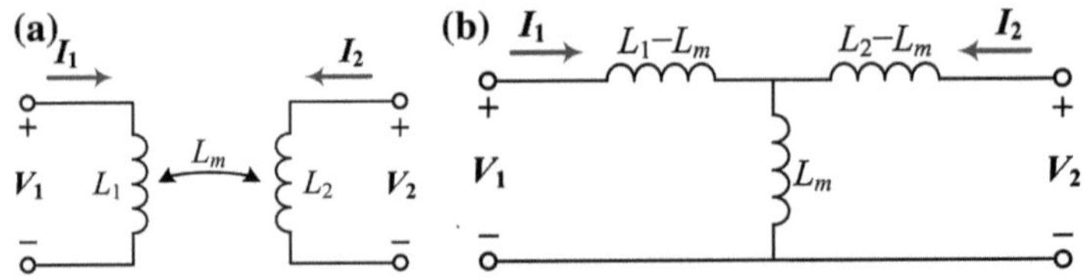

Fig. B.2 **a** Equivalent circuit model of inductive coupling between two inductors by a mutual inductance and **b** transformed equivalent T-circuit

In time-harmonic regime, the currents and voltages at the ports are given by

$$V_1 = j\omega L_1 I_1 + j\omega L_m I_2, \tag{B.4a}$$

$$V_2 = j\omega L_2 I_2 + j\omega L_m I_1. \tag{B.4b}$$

An equivalent T-circuit model can be obtained by connecting the lower terminals, as shown in Fig. B.2b [1–3].

References

1. J.-S. Hong, M.J. Lancaster, *Microstrip Filters for RF/Microwave Applications* (Wiley, New York, 2001)
2. J. Hong, Couplings of asynchronously tuned coupled microwave resonators. *IEE Proc. Microw. Antennas Propag.* **147**(5), 354–358 (2000)
3. W. Hayt, J. Kemmerly, S. Durbin, *Engineering Circuit Analysis* (McGraw-Hill, New York, 2011)

Appendix C
Even/Odd-Mode S Parameters

Figure C.1a illustrates a four-port network that is assumed to be symmetric and reciprocal. By applying the *even/odd-mode analysis*, the analysis of symmetric N-port networks is reduced to analyzing two $N/2$-port networks, namely, the even- and odd-mode networks (we restrict the analysis here to $N = 4$) [1–3].

In this method, two modes of excitation are defined, an even mode (symmetric mode) and an odd mode (antisymmetric mode). An even-mode excitation applies in-phase signals with the same amplitude to symmetrical ports (e.g. 1 and 3). Contrarily, an odd-mode excitation is that where anti-phase signals with the same amplitude are applied. Therefore, the whole network can be bisected along the symmetry plane into two identical halves. As shown in Fig. C.1b, the midplane is replaced with a magnetic wall (open circuit) for the even mode and with an electric wall (short circuit) for odd mode. Two two-port networks, namely, the even- and odd-mode networks, result from each of the halves of the symmetric four-port network (e.g. between ports 1 and 2) where the symmetry plane is replaced with a magnetic wall or an electric wall, respectively. The even- and odd-mode networks are characterized by

$$\boldsymbol{S_e} = \begin{pmatrix} S_{e11} & S_{e21} \\ S_{e21} & S_{e22} \end{pmatrix}, \tag{C.1a}$$

$$\boldsymbol{S_o} = \begin{pmatrix} S_{o11} & S_{o21} \\ S_{o21} & S_{o22} \end{pmatrix}, \tag{C.1b}$$

where $\boldsymbol{S_e}$ and $\boldsymbol{S_o}$ denote the even- and odd-mode scattering (S) matrices, respectively.

As is well known, an arbitrary excitation can be treated as a superposition of appropriate amplitudes of even and odd modes. Consequently, the order-4 S matrix of symmetric four-port networks can be obtained in terms of the even- and odd-mode S parameters as follows [1]

J. Naqui, *Symmetry Properties in Transmission Lines Loaded with Electrically Small Resonators*, Springer Theses,
DOI 10.1007/978-3-319-24566-9

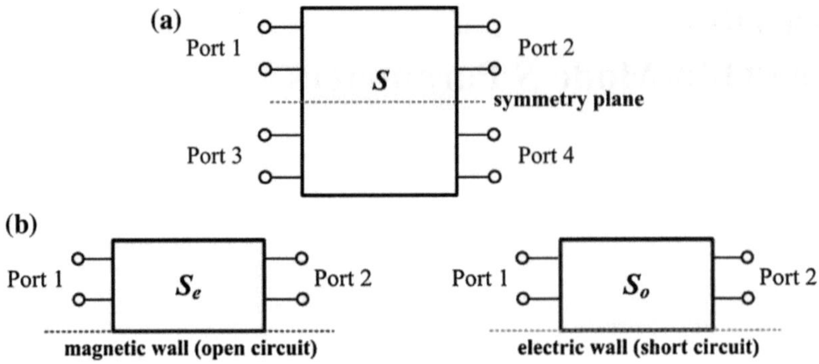

Fig. C.1 **a** Symmetric four-port network characterized by an order-4 S matrix. **b** Two-port even- and odd-mode networks characterized by order-2 S matrices

$$\boldsymbol{S} = \begin{pmatrix} S_{11} & S_{12} & S_{13} & S_{14} \\ S_{21} & S_{22} & S_{23} & S_{24} \\ S_{31} & S_{32} & S_{33} & S_{34} \\ S_{41} & S_{42} & S_{34} & S_{44} \end{pmatrix} = \begin{pmatrix} \boldsymbol{S_A} & \boldsymbol{S_B} \\ \boldsymbol{S_B} & \boldsymbol{S_A} \end{pmatrix}, \tag{C.2}$$

where

$$\boldsymbol{S_A} = \frac{1}{2}(\boldsymbol{S_e} + \boldsymbol{S_o}), \tag{C.3a}$$

$$\boldsymbol{S_B} = \frac{1}{2}(\boldsymbol{S_e} - \boldsymbol{S_o}). \tag{C.3b}$$

Conversely, the S parameters for the even and odd networks can be inferred from the S parameters for the whole network by

$$\boldsymbol{S_e} = \boldsymbol{S_A} + \boldsymbol{S_B}, \tag{C.4a}$$

$$\boldsymbol{S_o} = \boldsymbol{S_A} - \boldsymbol{S_B}. \tag{C.4b}$$

References

1. R.K. Mongia, J. Hong, P. Bhartia, I.J. Bahl, *RF and Microwave Coupled-Line Circuits* (Artech House, Boston, 1999)
2. J.-S. Hong, M.J. Lancaster, *Microstrip Filters for RF/Microwave Applications* (Wiley, New York, 2001)
3. D.M. Pozar, *Microwave Engineering*, 3rd edn. (Wiley, New York, 2005)

Appendix D
Mixed-Mode S Parameters

For the characterization of differential circuits, *mixed-mode scattering* (S) *parameters* are usually preferred over standard *single-ended S parameters*. Single-ended parameters measure single-ended signals at ports where one of the two terminals is referenced to ground. By contrast, mixed-mode parameters are defined at pairs of single-ended ports, known as *composite* or *mixed-mode ports* [1–3]. Signals at composite ports are decomposed into differential- and common-mode portions, and the simultaneous propagation of such modes is referred to as mixed-mode propagation.

Hence, a differential network may be viewed as a circuit of an even number of N single-ended ports or $N/2$ composite ports (see Fig. D.1). Some vector network analyzers (VNAs) are able to measure directly mixed-mode S parameters. Otherwise, these parameters may be obtained from the measurement of single-ended parameters, and by transformation expressions. For $N = 4$, the mixed-mode parameters defined as [1–3]

$$\boldsymbol{S_{mm}} = \begin{pmatrix} \boldsymbol{S_{dd}} & \boldsymbol{S_{dc}} \\ \boldsymbol{S_{cd}} & \boldsymbol{S_{cc}} \end{pmatrix} = \begin{pmatrix} S_{dd11} & S_{dd12} & S_{dc11} & S_{dc12} \\ S_{dd21} & S_{dd22} & S_{dc21} & S_{dc22} \\ S_{cd11} & S_{cd12} & S_{cc11} & S_{cc12} \\ S_{cd21} & S_{cd22} & S_{cc21} & S_{cc22} \end{pmatrix}, \tag{D.1}$$

can be expressed in terms of the single-ended parameters by

$$\boldsymbol{S_{dd}} = \frac{1}{2}\begin{pmatrix} S_{11} - S_{13} - S_{31} + S_{33} & S_{12} - S_{14} - S_{32} + S_{34} \\ S_{21} - S_{23} - S_{41} + S_{43} & S_{22} - S_{24} - S_{42} + S_{44} \end{pmatrix}, \tag{D.2a}$$

$$\boldsymbol{S_{cc}} = \frac{1}{2}\begin{pmatrix} S_{11} + S_{13} + S_{31} + S_{33} & S_{12} + S_{14} + S_{32} + S_{34} \\ S_{21} + S_{23} + S_{41} + S_{43} & S_{22} + S_{24} + S_{42} + S_{44} \end{pmatrix}, \tag{D.2b}$$

$$\boldsymbol{S_{cd}} = \frac{1}{2}\begin{pmatrix} S_{11} - S_{13} + S_{31} - S_{33} & S_{12} - S_{14} + S_{32} - S_{34} \\ S_{21} - S_{23} + S_{41} - S_{43} & S_{22} - S_{24} + S_{42} - S_{44} \end{pmatrix}, \tag{D.2c}$$

$$\boldsymbol{S_{dc}} = \frac{1}{2}\begin{pmatrix} S_{11} + S_{13} - S_{31} - S_{33} & S_{12} + S_{14} - S_{32} - S_{34} \\ S_{21} + S_{23} - S_{41} - S_{43} & S_{22} + S_{24} - S_{42} - S_{44} \end{pmatrix}. \tag{D.2d}$$

J. Naqui, *Symmetry Properties in Transmission Lines Loaded with Electrically Small Resonators*, Springer Theses,
DOI 10.1007/978-3-319-24566-9

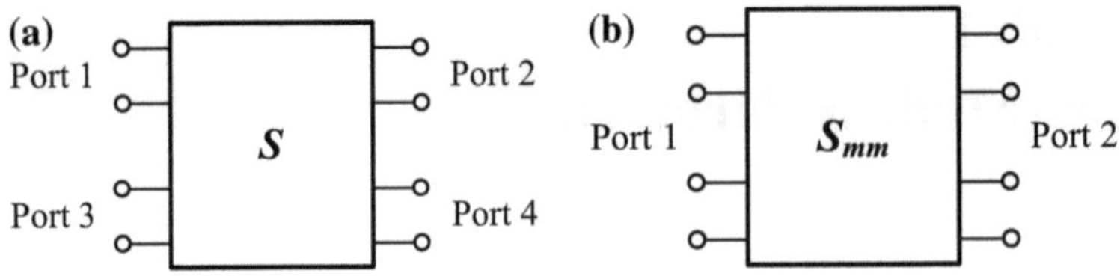

Fig. D.1 Port configuration for the transformation between **a** single-ended and **b** mixed-mode S parameters. The case with $N = 4$ corresponding to **a** four single-ended ports and **b** two composite ports is illustrated

$\boldsymbol{S_{dd}}$ stands for the differential-mode S parameters, $\boldsymbol{S_{cc}}$ represents the common-mode S parameters, and $\boldsymbol{S_{cd}}$ and $\boldsymbol{S_{dc}}$ account for the mode-conversion or cross-mode S parameters.

If the network is symmetric[1] and reciprocal, the cross-mode parameters are zero, whereas the differential- and common-mode parameters are the same as the odd- and even-mode single-ended parameters (Appendix C), i.e.

$$\boldsymbol{S_{dd}} = \boldsymbol{S_o}, \tag{D.3a}$$

$$\boldsymbol{S_{cc}} = \boldsymbol{S_e}. \tag{D.3b}$$

However, it should be highlighted that the reference impedance at the ports for the differential- and common-mode are, respectively,

$$Z_{0dd} = 2Z_0, \tag{D.4a}$$

$$Z_{0cc} = \frac{Z_0}{2}, \tag{D.4b}$$

where Z_0 is the reference impedance to calculate the single-ended S parameters. The previous relationships come from the definition of voltages and currents at the ports (not included in the Appendix). Therefore, if $Z_0 = 50\ \Omega$ as usual, the reference impedances to calculate the mixed-mode S parameters become $Z_{0dd} = 100$ and $Z_{0cc} = 25\ \Omega$.

References

1. D. Bockelman, W. Eisenstadt, Combined differential and common-mode scattering parameters: theory and simulation. IEEE Trans. Microw. Theory Tech. **43**(7), 1530–1539 (1995)
2. W.R. Eisenstadt, *Microwave Differential Circuit Design using Mixed-Mode S-Parameters* (Artech House, Boston, 2006)
3. G.D. Vendelin, A.M. Pavio, U.L. Rohde, *Microwave Circuit Design Using Linear and Nonlinear Techniques* (Wiley, New York, 2005)

[1] Ideally, a differential circuit should be perfectly symmetric.

Appendix E
Rejection Bandwidth Definition

This appendix discusses the relation between the stopband bandwidth inferred from the dispersion relation and the $-20\,\mathrm{dB}$ bandwidth obtained from the S parameters in a network with a significant number of cells. Particularly, the analysis is carried out by using the even-mode equivalent circuit model of a specific CSRR-loaded differential line (see Fig. E.1).

The dispersion relation of the considered network is plotted in Fig. E.2. In the circuit under study, the forbidden band includes the region where complex modes are present (between 1.42 and 1.61 GHz), plus an additional region in which the modes are evanescent (from 1.61 to 1.93 GHz). As is well known, the dispersion relation provides the frequency regions where signal propagation is allowed or forbidden for an infinitely periodic structure. For a finite number of cells, the most relevant effects that may occur are impedance mismatch in the allowed bands or small rejection within the forbidden bands. Therefore, an analysis to determine the rejection level that properly defines the forbidden band in a truncated structure is very convenient.

It is also well known that the transmission coefficient in two-port periodic structures suffers from ripple in the allowed bands. Such ripple is caused by impedance mismatch, and maximum transmission is produced at those frequencies where either the Bloch impedance [1, 2] is matched to the impedance of the ports (impedance matching) or the electrical length is a multiple of π (phase matching). Between adjacent transmission peaks, the transmission coefficient is a minimum at those frequencies where the phase is an odd multiple of $\pi/2$. The transmission coefficient at these frequencies of minimum transmission can be easily obtained according to [3]

$$f_{env}(\omega) = 10\log_{10}\left\{1 - \left|\frac{Z_B^2(\omega) - Z_0^2}{Z_B^2(\omega) + Z_0^2}\right|^2\right\}, \tag{E.1}$$

where $Z_B(\omega)$ is the Bloch impedance and Z_0 the reference impedance of the ports ($Z_0 = 50\,\Omega$). Expression (E.1) corresponds to the envelope of the transmission

J. Naqui, *Symmetry Properties in Transmission Lines Loaded with Electrically Small Resonators*, Springer Theses,
DOI 10.1007/978-3-319-24566-9

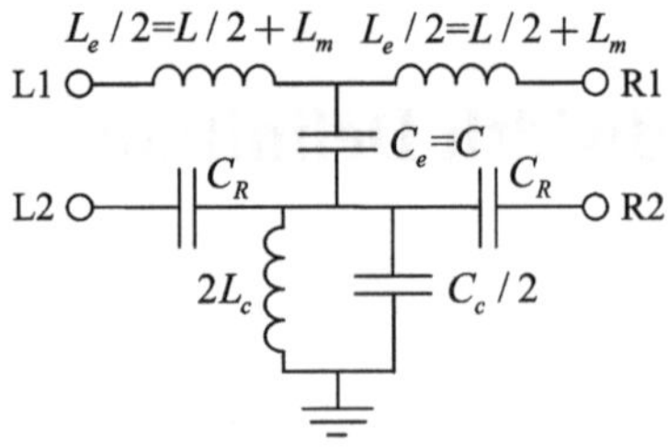

Fig. E.1 Even-mode circuit model (unit cell) of a CSRR-loaded differential line. The considered circuit parameters are: $L_e = 6.3\,\text{nH}$, $C_e = 1.1\,\text{pF}$, $L_c = 2.1\,\text{nH}$, $C_c = 3.2$, and $C_R = 0.1\,\text{pF}$ (those corresponding to the CSRR-based common-mode filter with $f_z = 1.5\,\text{GHz}$ in Table 5.1)

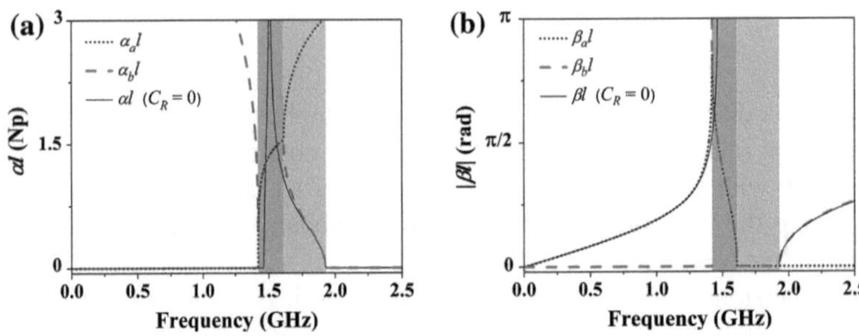

Fig. E.2 **a** Real part and **b** magnitude of the imaginary part of γl corresponding to the two modes of the considered network, where α is the attenuation constant, β the phase constant, and l the elemental cell period. If inter-resonator coupling is canceled ($C_R = 0$), the two modes degenerate into a single mode, also depicted. The evanescent and complex mode regions are highlighted in *light* and *dark gray*, respectively. Reprinted with permission from [5]

coefficient in the allowed bands of a two-port structure. However, the structure under analysis is described by a circuit model with two input and output ports (see Fig. E.1). In a two-port structure, the Bloch impedance is obtained by the ratio between the eigenvector variables (voltage and current). For the circuit under study, the pair of eigenvectors ($[V_{1a}, V_{2a}, I_{1a}, I_{2a}]$ and $[V_{1b}, V_{2b}, I_{1b}, I_{2b}]$ where a and b discriminate each mode) can be obtained by using

$$\begin{pmatrix} \boldsymbol{V_L} \\ \boldsymbol{I_L} \end{pmatrix} = \begin{pmatrix} \boldsymbol{A} & \boldsymbol{B} \\ \boldsymbol{C} & \boldsymbol{D} \end{pmatrix} \begin{pmatrix} \boldsymbol{V_R} \\ \boldsymbol{I_R} \end{pmatrix}, \tag{E.2}$$

and,

$$\cosh(\gamma l) = \frac{1}{2}\left(A_{11} + A_{22} \pm \sqrt{(A_{11} - A_{22})^2 + 4A_{12}A_{21}}\right), \tag{E.3}$$

and we can infer from them the characteristic impedance matrix as follows [4]

$$\boldsymbol{Z_c} = \boldsymbol{V_{ab}} \cdot \boldsymbol{I_{ab}}^{-1}, \tag{E.4}$$

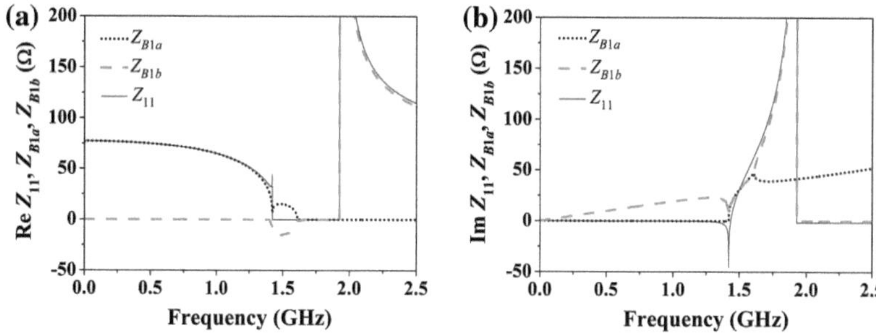

Fig. E.3 **a** Real and **b** imaginary part of Z_{11}, Z_{B1a}, and Z_{B1b}. Reprinted with permission from [5]

where $\boldsymbol{V_{ab}}$ and $\boldsymbol{I_{ab}}$ are 2×2 matrices formed by column vectors composed of the eigenvector voltages and currents

$$\boldsymbol{V_{ab}} = \begin{pmatrix} V_{1a} & V_{1b} \\ V_{2a} & V_{2b} \end{pmatrix}, \tag{E.5a}$$

$$\boldsymbol{I_{ab}} = \begin{pmatrix} I_{1a} & I_{1b} \\ I_{2a} & I_{2b} \end{pmatrix}. \tag{E.5b}$$

The diagonal elements of $\boldsymbol{Z_c}$ (Z_{11} and Z_{22}) are simply the ratios between the voltage and current in the corresponding port when the other port is left open-circuited, whereas the anti-diagonal elements determine the effects of the current injected into one port on the voltage in the other port when it is opened. Note that Z_{11} and Z_{22} are not the ratios of the voltage and current at ports 1 and 2 for each eigenvector. Let us denote such voltage to current ratios for port 1 by Z_{B1a} and Z_{B1b}. If we now compare Z_{11} to Z_{B1a} and Z_{B1b} (Fig. E.3), it can be appreciated that $Z_{11} \approx Z_{B1a}$ on the left of the forbidden band, whereas $Z_{11} \approx Z_{B1b}$ on the right of that band. This means that, in the allowed bands, the propagation in an infinite structure resulting by cascading the networks of Fig. E.1 with port L2 of the first cell left opened, can be described as if it was a two-port network that supports the eigenvector composed of the variables referred to port 1 (mode a and b on the left and right of the stop band, respectively) and Bloch impedance given by Z_{11}. This approximation is valid as long as mode mixing is negligible, as occurs in the whole allowed frequency region except in a narrow band on the left of the stop band, where modes a and b co-exist (this leads to a complex value of Z_{11} in that region).

From the previous statements, it follows that the envelope function of the transmission coefficient of any finite structure with ports L2 and R2 of the first and last cell, respectively, left open-circuited (actual conditions), can be approximated by means of (E.1), considering $Z_B = Z_{11}$. The envelope function in the allowed bands inferred through (E.1) with Z_{11} is compared to the response of a 12-cell structure

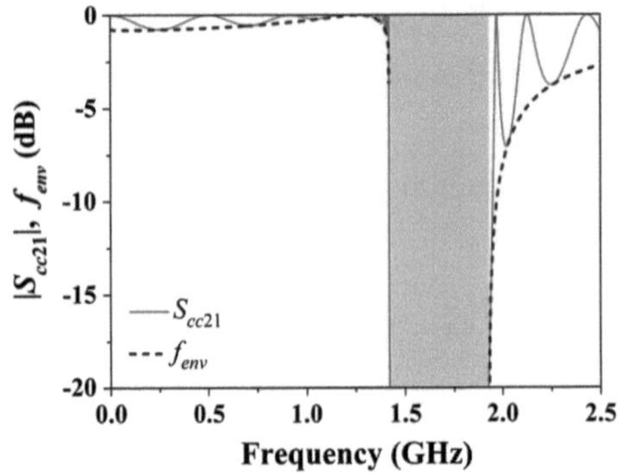

Fig. E.4 Transmission coefficient of the 12-cell structure resulting by cascading the 4-port network of Fig. E.1 with L2 and R2 ports of stages 1 and 12, respectively, left opened, and envelope function in the allowed bands. The forbidden band is depicted in *gray*. Reprinted with permission from [5]

in Fig. E.4. It is confirmed that we can use expression (E.1) to obtain the envelope function to a very good approximation.[2]

With regard to the stopband bandwidth, it coincides with the region delimited by those frequencies where the envelope function exhibits a vertical slope. This vertical slope is clearly achieved at a rejection level of 20 dB. In view of this result, the 20-dB rejection level is intimately related to the stopband bandwidth obtained by the dispersion relation, and is a good rejection level to define the common-mode rejection bandwidth.

References

1. D.M. Pozar, *Microwave Engineering*, 3rd edn. (Wiley, New York, 2005)
2. F. Martín, *Artificial Transmission Lines for RF and Microwave Applications* (Wiley, New York, 2015)
3. M. Gil, J. Bonache, I. Gil, J. García-García, F.Martín, On the transmission properties of left-handed microstrip lines implemented by complementary split rings resonators. Int. J. Numer. Model. Electron. Networks Devices Fields **19**(2), 87–103 (2006)
4. R.K. Mongia, J. Hong, P. Bhartia, I.J. Bahl, *RF and Microwave Coupled-Line Circuits* (Artech House, Boston, 1999)
5. J. Naqui, A. Fernández-Prieto, M. Durán-Sindreu, F. Mesa, J. Martel, F. Medina, F. Martín, Common-mode suppression in microstrip differential lines by means of complementary split ring resonators: theory and applications. IEEE Trans. Microw. Theory Tech. **60**(10), 3023–3034 (2012)

[2]It should be emphasized that the proposed procedure to infer the envelope function is approximate. However, expression (E.1) provides very good results because the coupling capacitance between resonators, C_R, is small in the considered structures. If this condition is not fulfilled, mode mixing is more pronounced, and the envelope function that results when the ports L2 and R2 of the input and output cells, respectively, are opened, cannot be inferred following this simple procedure.

Appendix F
Envelope Detector Design

An envelope detector as that depicted in Fig. 6.25a works on a charging-discharging scheme [1]. Assuming an ideal diode, in the positive half-cycle of the modulated signal, the diode is forward-biased and the capacitor C charges rapidly to the peak value of the modulated signal with a time constant

$$\tau_c = Z_0 C << T_c, \tag{F.1}$$

where $T_c = 1/f_c$ is the period of the carrier signal. When the input signal falls below the peak value, the diode becomes reverse-biased, and the capacitor discharges slowly through the load resistor R until the next positive half-cycle, with a time constant, τ_d, satisfying

$$T_c << \tau_d = RC << T_m. \tag{F.2}$$

In the readout circuit in Fig. 6.25a, at the detector output, the envelope signal in time domain can be visualized by an oscilloscope. The waveform of the envelope is directly related to the transfer function (ideally they should be identical). Ideally, the input impedance of an oscilloscope should be infinite. However, in practice, loading effects arise (the input impedance is finite and complex, with resistive and capacitive components in parallel). Thus, for low loading effects, high resistance and low capacitance are required. For the available oscilloscope the input impedance is 1 MΩ in parallel with 14 pF [2]. Regardless of the value of C, the input capacitance of the oscilloscope is not small enough, and condition (F.1) cannot be satisfied. Nevertheless, note that the input impedance of the oscilloscope is indeed an RC filter as that of the detector circuit. Therefore, we may take advantage of the loading effects to inherently implement the filter with the measurement. However, as mentioned before, the input capacitance is not small enough. For this reason, a single-ended active probe is used, achieving an input impedance of 1 MΩ in parallel with 1 pF. Thereby conditions (F.1) and (F.2) are satisfied. For the sensor in Fig. 6.24, the conditions yield

J. Naqui, *Symmetry Properties in Transmission Lines Loaded with Electrically Small Resonators*, Springer Theses,
DOI 10.1007/978-3-319-24566-9

$$T_c/\tau_c = 13.2 >> 1, \tag{F.3a}$$

$$T_m/\tau_d = 5 \cdot 10^5 (60\ \text{rpm}) >> 1, \tag{F.3b}$$

$$T_m/\tau_d = 10^4 (3{,}000\ \text{rpm}) >> 1, \tag{F.3c}$$

$$\tau_d/T_c = 1.52 \cdot 10^3 >> 1. \tag{F.3d}$$

References

1. S. Haykin, *Communication Systems* (Wiley, New York, 2001)
2. J. Naqui, F. Martín, Transmission lines loaded with bisymmetric resonators and their application to angular displacement and velocity sensors. IEEE Trans. Microw. Theory Tech. **61**(12), 4700–4713 (2013)

MIX
Papier aus verantwortungsvollen Quellen
Paper from responsible sources
FSC® C105338

If you have any concerns about our products,
you can contact us on
ProductSafety@springernature.com

In case Publisher is established outside the EU,
the EU authorized representative is:
Springer Nature Customer Service Center GmbH
Europaplatz 3, 69115 Heidelberg, Germany

Printed by Libri Plureos GmbH
in Hamburg, Germany